STP 998

Electrical Insulating Oils

Herbert G. Erdman, editor

ASTM
1916 Race Street
Philadelphia, PA 19103

ASTM Publication Code Number (PCN): 04-998000-21
ISBN: 0-8031-1179-7

NOTE

The Society is not responsible, as a body,
for the statements and opinions
advanced in this publication.

Peer Review Policy

Each paper published in this volume was evaluated by three peer reviewers. The authors addressed all of the reviewers' comments to the satisfaction of both the technical editor(s) and the ASTM Committee on Publications.

The quality of the papers in this publication reflects not only the obvious efforts of the authors and the technical editor(s), but also the work of these peer reviewers. The ASTM Committee on Publications acknowledges with appreciation their dedication and contribution of time and effort on behalf of ASTM.

Printed in Ann Arbor, MI
September 1988
Second Printing, Ann Arbor, MI
May 1992

Foreword

This publication, *Electrical Insulating Oils,* contains papers presented at the symposium of the same name held in Bal Harbour, Florida on 19–20 Oct. 1987. The symposium was sponsored by ASTM Committee D-27 on Electrical Insulating Liquids and Gases. Herbert G. Erdman, PSE&G Research Corp., Maplewood, NJ, presided as symposium chairman and was editor of this publication.

Contents

Overview

This symposium was organized to update information on the important role of petroleum-based oil as an insulating and heat transfer agent in electrical apparatus.

The properties of these insulating oils are of vital importance to the service life of electrical equipment and have become increasingly important as operating voltages have increased to 500 kV and above and internal equipment spacings have decreased. The papers in this symposium have been directed towards the changing requirements of the insulating media as demanded by the changing design of the equipment, and they are of particular value to three main groups: the refiners of the oil; the manufacturers of the electrical equipment; and the end users (principally the electric utilities).

The papers in this book address the factors involved in the manufacture and use of the product that will meet these stringent requirements.

From the refiners' standpoint, the papers discuss how the crude oil is selected and what must be done to produce an end product with high dielectric strength, heat transfer capability, low-temperature pour point, resistance to oxidation, and long service life. These factors determine the specifications of the product.

From the users' standpoint, the papers discuss the significance of these specifications and test methods to determine if the specifications have been met, as well as tests to check the condition of oil after use in the equipment. Included is the very important diagnostic test for electric equipment in service, the analysis of dissolved gases in the oil.

Of special interest is section 5, which is devoted to a relatively new problem caused by electrostatic charges built up in the oil due to forced cooling in large transformers.

This volume is of value to all who refine or use insulating oils because it addresses the latest technologies involved.

The first section of this book, "Refining and Specification Limits," addresses methods used in refining crude oil to produce a product which is suitable for both electric insulation and heat transfer. Included is the problem of handling and shipping the product once produced. Contamination can occur from tank cars and drums that are not completely devoid of foreign materials or not properly sealed. The paper by Manger, which discusses cross contamination with polychlorinated biphenyls (PCB), is included to describe a specific problem. This paper updates the local and federal regulations dealing with PCB contamination.

The second section, "Significance of Application," addresses the reasons behind the various specifications for insulating oils and the significance of these specifications to the users.

The third section, "Analysis of Oil," discusses the methods of analysis to determine if these specifications have been met. There are two conditions to be considered: (1) analyzing new oil as shipped; and, perhaps more important, (2) the analysis of oil after certain periods of time in the electrical equipment to determine deterioration and predict further usefulness of the oil.

The fourth section, "Dissolved Gas in Oil," discusses the analysis of gases dissolved in oil and the significance of various gases to the condition of the electrical equipment involved. This is a very useful diagnostic test, particularly to monitor the performance of large high-voltage transformers.

The fifth section, "Electrostatic Buildup in Transformer Oil," perhaps one of the most provocative, has to do with a recent problem that has arisen in large transformers where the pumping of the oil is required to limit the temperature rise in the transformers. Pumping a liquid with high dielectric strength and very low quantities of moisture over various materials in the pump

1

and in the insulating material, in many cases, has caused an electrostatic charge buildup in the oil and has resulted in electrical failure in several large transformers throughout the country. The papers in this section present the latest information on this phenomena. There is much more to be learned of this complex process, and much work is being done throughout the country to try and solve this problem.

The papers in this book should provide the reader with the latest technologies in selecting and handling this very important insulating media for electric equipment. This is a very significant area for equipment designers. Large capacity size and weight of electrical equipment is of vital importance, resulting in smaller internal spacings requiring the best dielectric material available.

ASTM Committee D27 on Electrical Insulating Liquids and Gases is continually working to update this material, improve test procedures, and investigate newly refined products that appear on the market. Refiners are continuously producing new insulating oils from petroleum crude oil. In addition, manufacturers are continually producing synthetic insulating fluids. Committee D27 on Electrical Insulating Liquids and Gases addresses these new products as they come on the market.

Herbert G. Erdman,

PSE & G Research Corp.; Maplewood, NJ 07040; symposium chairman and editor

Section I—Refining and Specification Limits

Thomas G. Lipscomb, II[1]

Mineral Insulating Oil Manufacture and Safekeeping

REFERENCE: Lipscomb, T. G. II, **"Mineral Insulating Oil Manufacture and Safekeeping,"** *Electrical Insulating Oils, STP 998,* H. G. Erdman, Ed., American Society for Testing and Materials, Philadelphia, 1988, pp. 5–24.

ABSTRACT: This paper presents a general overview of the methods used to refine conventional mineral insulating oils. The discussion will outline the individual refining processes and the effect each may have on the composition of the oil. Included will be a brief discussion of the Occupational Safety and Health Administration's Hazard Communication Standard, 29 CFR 1910.1200 as it can apply to mineral insulating oil and some suggestions for handling and storage of the oil to reduce contamination and degradation.

KEY WORDS: transformer oil, mineral insulating oil, refining, naphthenic transformer oil, paraffinic transformer oil, storage, handling, cancer hazard warning, OSHA hazard communication standard

In planning a seminar on the subject of insulating oil of petroleum origin, it was recognized that many people involved in using "mineral insulating oils" or "transformer oils" have probably never known how they are made. Only those persons involved in their manufacture or in research on the subject have direct knowledge.

The author was asked to provide information of a nonproprietary nature on the general subject to serve as a basis for understanding some of the properties of transformer oils that will be discussed in subsequent papers.

Discussion

Refining

Mineral transformer oil today is a compromise between what is needed by the equipment, what is desired by the customer, and what can be achieved without overwhelming complexity in the refining process. This compromise is well exemplified by ASTM Specification for Mineral Insulating Oil Used in Electrical Apparatus (D 3487). The need to meet this specification in the most efficient manner is the aim of the refiner.

The manufacture of transformer oil begins with a choice of crude oil type: naphthenic, paraffinic, or mixed base. Convention in the United States has been to use naphthenic crudes, since they generally contain very much lower concentrations of n-paraffins (wax), and the resultant product has a low pour point without dewaxing being required. This convention at one time applied everywhere, but necessity has driven some regions of the world to use mixed-base or paraffinic crudes as sources. All useable lube crudes can be manipulated by suitable combinations of processes to result in a useable transformer oil. The degree of difficulty in refining and the product yield will vary with the crude.

There is also another little-appreciated benefit of using naphthenic oils; their viscosities change much faster with temperature than those of paraffinic-derived oils. Normally, this might

[1]Technical advisor, Exxon Company, U.S.A., Houston, TX 77252.

be considered a detriment, but not in this case. As the temperature rises in a transformer, the oil becomes less viscous and the heat transfer rate is improved. For oils of equal viscosity at 40°C, the heat transfer coefficient can be as much as 8 to 11% greater at 100°C for an oil of 30 VI (viscosity index) versus an oil of 100 VI.

Since the world supply of good "low cold test" naphthenic crudes has declined, we have seen paraffinic crudes being used as sources, particularly in South America, Mexico, and Europe. One such oil was commercially introduced in the United States. At the time this paper was written, it was withdrawn from the market.

The definition of a naphthenic crude by the U.S. Bureau of Mines contains no mention of the pour point of the distillate nor of the n-paraffin content. This classification system is shown in Fig. 1. By this system it is possible to have nine different crude types, depending upon the various combinations of light key and heavy key classifications.

ASTM Committee D27 was more pragmatic and chose to use the definitions in ASTM Definitions of Terms Relating to Electrical Insulating Liquids and Gases (D 2864) as shown in Fig. 2. Although not technically correct, they represent the main concerns about pour points, low-temperature viscosities, and the means needed to achieve the desired properties.

Refining processes are used to eliminate undesirable compounds and to retain those that are desirable. The elimination procedure may be by simple removal or by conversion to more desirable compounds. Both types of elimination procedures are represented by refining processes commonly in use today.

Crude Distillation

The first step in producing a transformer oil is distillation of the crude to produce a suitable distillate feedstock. This is usually done on a pipe still. The distillate can be taken as a side stream product from either the atmospheric tower or the vacuum tower. Or a very broad stream can be taken, processed partially, and then redistilled to the desired narrow boiling range. Figure 3 shows a schematic flow diagram of a typical pipe still operation.

In this process, "light ends" are removed in a prefractionator tower (distillation column) not shown in the diagram, and "middle distillate" type products are removed in the atmospheric tower. The "reduced crude" from the bottom of the atmospheric tower then goes to the final, or vacuum, tower. This distillation column is run under vacuum to keep the boiling point of the individual fractions well below the cracking temperature, that is, the temperature at which thermal decomposition of the oil molecules would begin. Cracking of hydrocarbons usually becomes rapid enough to detect at about 370°C (700°F).

Steam is usually added to the bottom of the vacuum tower to further lower hydrocarbon partial pressure and reduce the temperature required for distillation still more. The transformer oil distillate, being relatively low boiling, is removed near the top of the tower. As mentioned earlier, the transformer oil distillate could also have been removed as a lower side stream on the atmospheric tower.

A question sometimes asked is, "What portion of crude can be made into transformer oil?"

	Gravity of Key Fraction, °API	
	Light Key Fraction (482-527°F)	Heavy Key Fraction (740-790°F)
Paraffinic	>40	>30
Intermediate	33-40	20-30
Naphthenic	<33	<20

FIG. 1—*Crude oil classifications (U.S. Bureau of Mines).*

Naphthenic Oil

A term applied to mineral insulating oil derived from special crudes having very low, naturally occuring n-paraffin (wax) contents. Such an oil has a low natural pour point and does not need to be dewaxed nor does it usually require the use of a pour depressant.

Paraffinic Oil

A term applied to mineral insulating oil derived from crudes having substantial contents of naturally occurring n-paraffins (wax). Such an oil must be dewaxed and may need the addition of a pour depressant in order to exhibit a low pour point.

FIG. 2—*Definitions of naphthenic oil and paraffinic oil.*

For most useable crudes, the distillate yield can be from 3 to 10% of the whole crude. There are two limiting factors. On the front end, the flash point limits the amount of light molecules that can be present. The second limitation is the maximum viscosity. As increasing amounts of heavier molecules are included, the viscosity of the whole fraction increases until the specified limit is reached.

There are other practical considerations. The molecules boiling above and below the desired fraction are valuable for other uses. Pipe stills are not perfect; there is overlap between fractions. If the maximum amount of transformer distillate is taken, then a shortage of the next lower and/or higher boiling fractions may develop.

In looking at some current typical transformer oils versus ASTM D 3487, none seems to be at the minimum flash point, and none is at the maximum viscosity, so compromise has won again.

Table 1 shows the boiling range of a distillate from a naphthenic crude and one from a paraffinic crude [by ASTM Test Method for Boiling Range Distribution of Petroleum Fractions by Gas Chromatography (D 2887)]. Both distillates were used to make finished transformer oil.

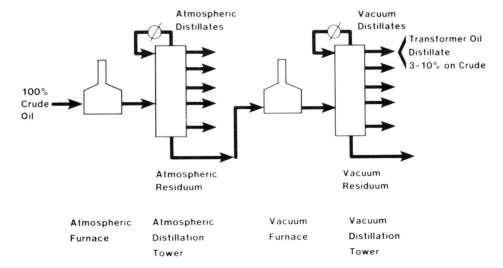

FIG. 3—*Distillation to make transformer oil base stock.*

TABLE 1—*Comparison of naphthenic and paraffinic distillates boiling ranges.*

	Naphthenic	Paraffinic
Initial boiling point, °C	283	239
5% distilled, °C	271	306
50% distilled, °C	330	372
95% distilled, °C	384	419
Final boiling point, °C	429	531

The data for the naphthenic distillate can be translated into an average molecular weight of 255, or an average molecule with 18 carbon atoms (carbon number = 18), if all of the compounds were paraffins. Similarly, the molecular weight of the paraffinic distillate would be about 300.

Not all of the compounds present in naphthenic distillates are paraffins, far from it. There is a great variety of compounds present, and, in general, the same types of compounds are present in all crudes. The difference in crudes can be found in the relative amounts of the various compound types and in the degree and kind of substitution on the ring structures.

Molecular Components of the Distillate

Before examining the next processing steps, we need to illustrate the various molecule types present in the distillate. Refining is essentially the effort to remove or preserve these compounds.

Paraffins, normal-, iso-, and cyclo-, are shown in Figs. 4 and 5. Remember, we said that the average carbon number was 18, yet the structures shown do not have 18 carbons. We are illustrating the compound types present, not the actual compounds themselves in order to simplify the drawings.

A naphthenic crude-derived distillate will have very low n-paraffin content. There will be some present, but the values will be so low as to be almost unmeasurable. Not all naphthenic

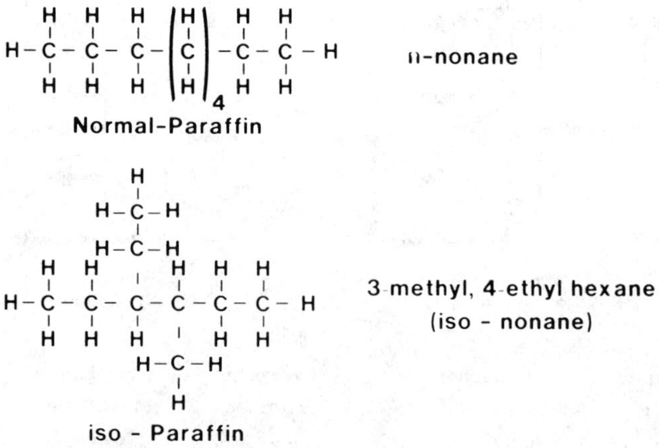

FIG. 4—*Paraffins.*

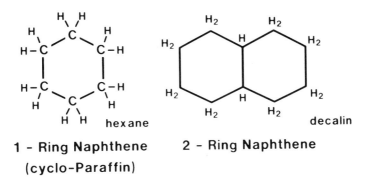

1 - Ring Naphthene   2 - Ring Naphthene

(cyclo-Paraffin)

FIG. 5—*Naphthenes.*

crudes are wax free. There are some which meet the U.S. Bureau of Mines definition, but which contain enough wax to make them unsuitable for "low cold test" products.

The naphthenic distillate will then contain considerable amounts of iso-paraffins and cyclo-paraffins. The iso-paraffins have very low crystallization temperatures, as do the cyclo-paraffins, provided that they do not have very long n-paraffin side chain substitutions. All these cyclo-paraffins, also called naphthenes (hence the name naphthenic crude), will have some substitution of side chains, probably several short or branched chains.

We want to keep the iso-paraffins and cyclo-paraffins and, if necessary, reject n-paraffins.

Olefins are shown in Fig. 6. The double bond makes them highly reactive to form sludges, acids, aldehydes, etc. There are several ways to eliminate them, which will be discussed later.

Aromatics are a major constituent of the distillate. Figure 7 does not show any side chain substitution, but at this boiling range all of the aromatics present will have side chains. The degree and kind of substitution strongly affects the viscosity/temperature properties of these aromatics. Note also mixed molecules, naphtheno-aromatics, where part is aromatic and part is naphthenic. Indane is an example.

Aromatics are both good and bad. We need to keep some but not all. In general, as the aromatic content is decreased, the gassing tendency of the oil increases, and the impulse

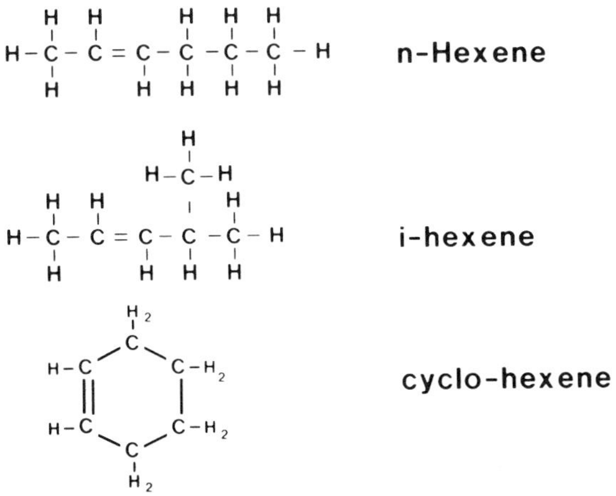

FIG. 6—*Olefins.*

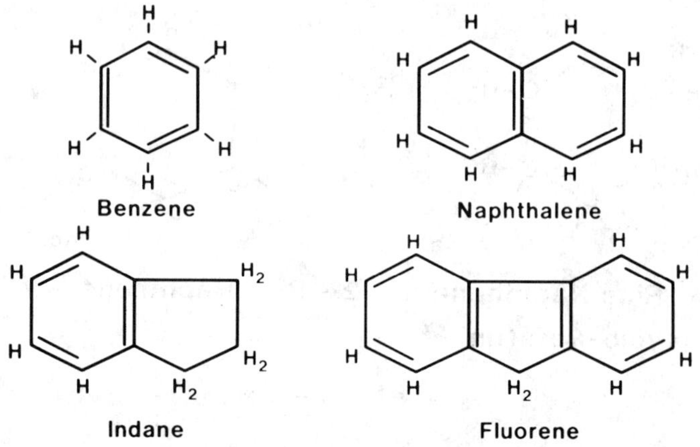

FIG. 7—*Aromatics.*

strength goes up. Also, some aromatics seem to be mild oxidation inhibitors. The refiner must then try to remove some aromatics, leave some aromatics, be selective about the process, and compromise between the desire for good antigassing properties, good impulse strength, and good oxidation resistance.

Polar compounds, and there are many types, are shown beginning in Fig. 8. Heterocyclic aromatics containing nitrogen or oxygen are generally undesirable. They are not stable as to oxidation, color, etc. Thiophenes may be an exception. Certain thiophenes are known to be moderate oxidation inhibitors.

Polar oxygenates shown in Fig. 9 are all undesirable. If they are not already, they soon will degrade further to become acids. As acids they contribute to corrosion, low interfacial tension, and poor dissipation factor.

Sulfur compounds, except for thiophene, are shown in Fig. 10. Some of these may be mild antioxidants, but they also react with copper and have objectionable odors. These compounds must be removed.

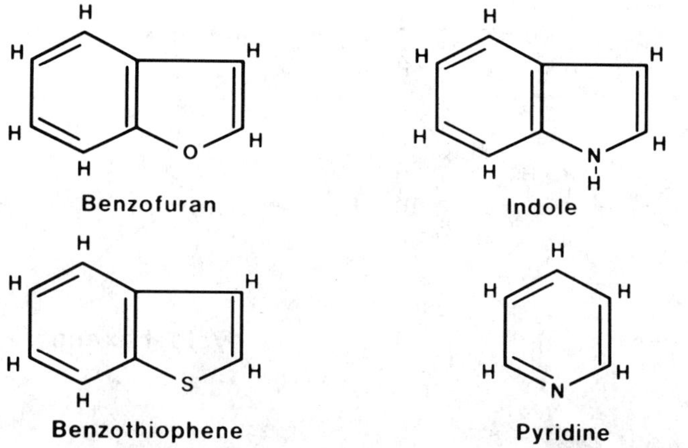

FIG. 8—*Polars, heterocyclics.*

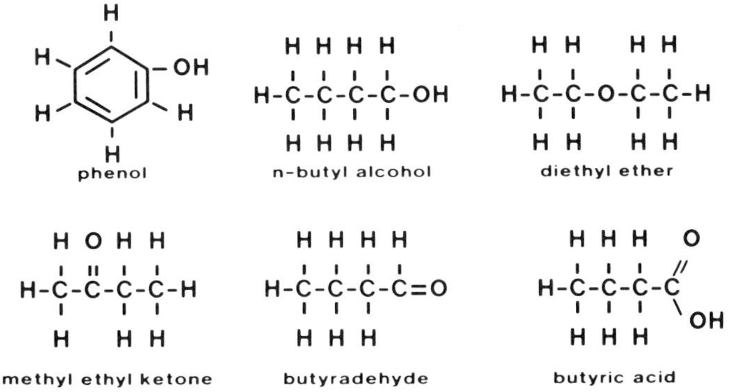

FIG. 9—*Polars, oxygenates.*

Actually, there are two more ever-present materials we must consider: water and particulates. These must be removed because they are as deleterious to the performance of the finished oil as any of the molecules previously mentioned.

Refining the Distillate

In this section we will describe each of the major refining processes. In a later section we will compare the physical and chemical effects of the processes on each molecule type present in the distillate. In examining the refining processes, it is well to keep in mind that it is rare to find a single process that will do all. Combinations of processes are the rule. We will list some of the more popular combinations, but not by any means all combinations.

Acid Treating—The first refining process is acid refining or acid treating. Many years ago it was the "one" accepted method, but in reality it was a dual process, for it was combined with clay treating. Figure 11 shows a block diagram of the process.

In this process, the transformer oil distillate is initially contacted with 90 to 99% sulfuric acid. A very light treat, 2 to 4 volume % acid based on the distillate, is used. The contact time of acid and distillate varies between about 0.1 and 4 s. Following mixing, the acid sludge that is formed from the chemical reaction of acid and oil is separated from the oil by centrifuging. At this point some harmful sulfonic acids and nitrogen bases are removed as sludge from the oil.

The oil from the centrifuging operation contains oil-soluble acidic compounds resulting from the reaction with sulfuric acid. Upon neutralization with sodium carbonate or caustic soda and

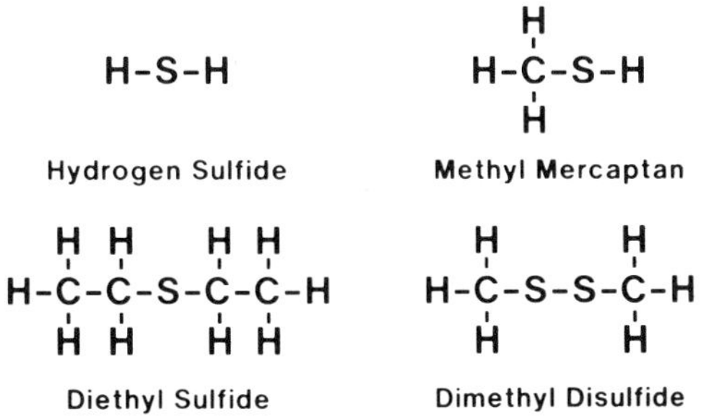

FIG. 10—*Sulfur compounds.*

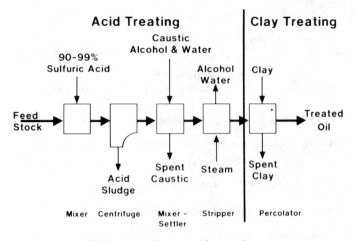

FIG. 11—*Acid treating/clay treating.*

extraction of the oil with alcohol, these acidic compounds are removed as acid salts. The neutralized oil is then stripped with steam to remove alcohol and water.

Processing of the oil to this stage may be described as a pretreatment operation, since the oil does not meet the electrical insulating oil criteria as outlined in ASTM D 3487. A finishing step, percolation through a bed of fine clay, is undertaken to produce an acid treated insulating oil of acceptable quality. The clay percolation step is necessary not only to remove compounds not reacted out by the acid treating, but also to remove residual traces of sodium sulfonates created by the neutralization of the sulfonic acids. The finished electrical insulating oil is neutral in acidity, noncorrosive, and has good electrical properties. More detail on the changes in chemical/physical properties that take place will be given later in this paper.

Hydrotreating—The next processes that will be described are hydrotreating and hydrogenation. The block flow diagrams for both processes are shown in Fig. 12. The two processes differ in severity of treatment of the oil. In hydrotreating (a relatively mild process), the aim is to remove unwanted sulfur, nitrogen, oxygen, and olefinic compounds without materially removing the aromatics present and to improve the odor, color, and color stability. For the purposes of this discussion, "hydrogenation" will mean a process that is so severe that a significant portion of the aromatics present is removed (converted) also.

In hydrotreating, hydrogen gas and the transformer oil distillate are chemically reacted over a fixed catalyst bed at elevated temperatures and pressures. In the hydrotreater, harmful material

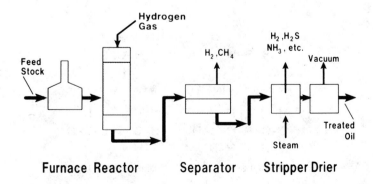

FIG. 12—*Hydrotreating/hydrogenation.*

(such as sulfur compounds, naphthenic acids, and nitrogen compounds) react with hydrogen. The products of these reactions are: (1) stable hydrocarbon compounds similar to naturally occurring hydrocarbon compounds; and (2) gases such as hydrogen sulfide, ammonia, and water. The conditions of pressure and temperature used in the hydrotreating process are not sufficient to cause cracking of the oil. The highest temperatures experienced by insulating oil occur in the pipe still during distillation.

After the hydrotreating step, the oil goes to a steam stripper where the gaseous reaction products (such as hydrogen sulfide, ammonia, and water) are removed. The oil can be dried at this stage by using a subsequent vacuum stripper.

The hydrotreated oil by itself is not usually a satisfactory product. The process is too mild to remove all deleterious compounds. Some other process such as clay percolation or solvent extraction must also be used to arrive at a suitable finished oil. Both combinations have been used by the industry.

Hydrogenation can differ from hydrotreatment in temperature and pressure, but usually also employs a different catalyst, one that is more reactive for the hydrogenation of aromatics to cyclo-paraffins.

There are commercially available a wide variety of catalysts used in hydrotreating and/or hydrogenation. The science of catalysis is now well enough advanced to permit the user to select a catalyst that will accentuate a specific reaction, such as desulfurization, dearomatization, denitrogenation, etc.

Hydrogenation can, by itself, serve to perform the functions of both solvent extraction (aromatics reduction) and hydrotreating (polar compound reduction), but it is so severe as to remove most naturally occurring antioxidants at the same time it removes oxidation promoters and unstable compounds. In this case, synthetic antioxidants must be used with the oil for the product to meet the oxidation requirements of ASTM D 3487.

Solvent Extraction—The solvent extraction block flow diagram is shown in Fig. 13. In this process the oil feed is contacted countercurrently in a treating tower or centrifugal extractor using an immiscible (or partly miscible) solvent that is more dense than the oil and is selective

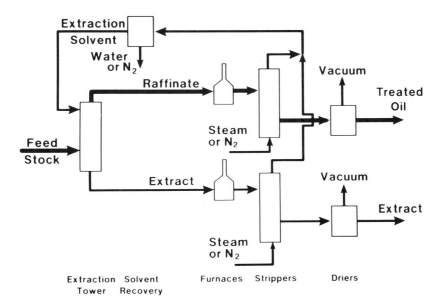

FIG. 13—*Solvent extraction.*

for aromatics. The extracted oil (now called raffinate) containing some dissolved solvent goes to a stripping system where the solvent is recovered and reused.

The bulk of the solvent from the treater tower containing some oil (now known as extract) goes to a separate stripping system where the solvent is removed from the extract for reuse. The extract produced has a high content of aromatics.

Solvents that have been used to treat transformer oil are: phenol, furfural, N-methyl-2-pyrrolidone, and liquid sulfur dioxide. All are selective for removal of aromatics, but the degree of selectivity will differ somewhat among the solvents and can even be changed by various means. It should be noted that none of these solvents is perfectly selective for aromatics. Desirable compounds are partly soluble, and so some are removed in the extract. Likewise, some of undesirable compounds are left behind. In general, all of the above solvents tend to remove a greater proportion of the more highly condensed ring aromatics first, the less condensed ring systems next, etc. Solvent extraction by itself is not a sufficient process. It must be used in combination with other processes.

Solvent Dewaxing—Dewaxing is not required if the crude source used is naphthenic (low cold test), but is required for oils made from paraffinic crudes or waxy naphthenic crudes.

Figure 14 is a block diagram of a solvent dewaxing unit. Waxy oil is mixed with a suitable antisolvent for wax (n-paraffins) and chilled to a low temperature to cause the n-paraffins to precipitate. The wax is then filtered out, usually on a rotary filter. The solvent is recovered from the dewaxed oil and wax for reuse.

Typical solvents that can be used in this process are liquid propane, methyl ethyl ketone, and methyl iso-butyl ketone. The ketones are especially good antisolvents for wax but typically must be used with another cosolvent to prevent rejection of desired molecules.

When using solvent dewaxing, the pour point of the dewaxed oil is usually limited to −18°C (0°F) to −23°C (−10°F) as a practical matter. It is theoretically possible to achieve a lower pour point, but most refinery equipment is not designed for very low temperature operation. Obviously, if a pour point of −40°C (−40°F) or lower is desired, either a further processing step is required, such as hydrodewaxing, or a pour depressant must be added to the oil. Pour depressants are in use in Europe but are not used in the United States.

Catalytic Dewaxing—A process capable of very low pour points is that of hydrodewaxing or catalytic dewaxing. Figure 15 gives a block diagram of this process. The equipment used is very similar to that used in hydrotreating. Only the catalyst needs to be different.

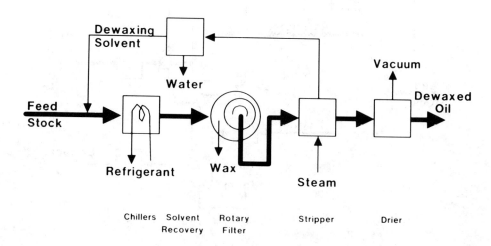

FIG. 14—*Solvent dewaxing.*

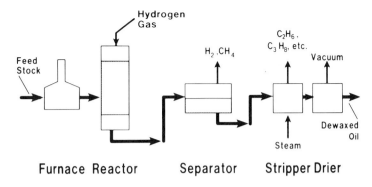

FIG. 15—*Catalytic dewaxing*

The dewaxing catalyst (for example, mordenite) can best be visualized as a hollow sphere with holes in the sides leading into the interior cavity. The hole size is such that a straight chain hydrocarbon (n-paraffin) can protrude through the hole into the interior, but a branched-chain (iso-paraffin) or cyclic structure cannot.

The waxy oil feed to the unit is mixed with hydrogen gas at elevated temperature and pressure and allowed to flow over the catalyst just as in hydrotreating. As a normal paraffin molecule protrudes into a side hole of the catalyst it is "cracked" off, typically in units of three carbons. The olefins formed immediately react with the hydrogen present to make saturated hydrocarbons. Instead of solid wax as a by-product, the wax is converted to light hydrocarbons, chiefly propane, by this process.

This process is capable of producing dewaxed oils with pour points of $-45°C$ ($-50°F$). No pour depressants are required.

Physical and Chemical Effects of Refining

Having discussed the molecular compound types present and the refining processes commonly in use, let us now see how the processes affect the molecular mixtures and, in turn, how that may affect the properties of the finished transformer oil.

Figure 16 shows that n-paraffins are removed only by the dewaxing processes.

Figure 17 shows that iso-paraffins and naphthenes are not affected by any process, except that small amounts may be solubilized into the extract phase in solvent extraction.

Figure 18 shows that olefins are removed by acid treating, while hydrotreating/hydrogenation converts them to saturated hydrocarbons which remain in the oil.

Acid Treating
Clay Treating
Hydrotreating   } **No Effect**
Hydrogenation
Solvent Extraction

$$\underline{\textbf{Solvent Dewaxing}}$$
$$CH_3(CH_2)_n CH_3 \xrightarrow{\text{Solvent}} \textbf{Removal By Filtering}$$

$$\underline{\textbf{Hydro-Dewaxing}}$$
$$CH_3(CH_2)_n CH_3 + H_2 \xrightarrow{\text{Cat.}} C_3H_8, \textbf{Etc.}$$

FIG. 16—*Hydrocarbon reactions: n-paraffins.*

Acid Treating
Clay Treating
Hydrotreating
Hydrogenation   No Effect
Solvent Extraction*
Solvent Dewaxing
Hydro-Dewaxing

*Small Amounts May Be Solubilized Into The Extract Phase.

FIG. 17—*Hydrocarbon reactions: iso-paraffins, naphthenes.*

Figures 19 and 20 show the slight removal of highly condensed aromatics by acid treating under the mild conditions used for transformer oil. Hydrotreating converts some of them to partially saturated hydrocarbons. Depending upon the severity of the process, hydrogenation can fully saturate the aromatics, converting them to naphthenes. Solvent extraction will remove substantially all of the highly condensed aromatics. Clay treating, solvent dewaxing, and hydro-dewaxing have no effect. It is desirable to leave a substantial amount of aromatics (of the right kind) in the oil to balance the impulse strength, gassing tendency, and oxidation resistance of the finished oil. This is achieved by adjusting the severity of the processes that affect aromatics content.

Figures 21 and 22 illustrate the reactions with benzothiophenes. Acid-treating can remove them; hydrotreating may convert a limited amount, and hydrogenation can totally convert them if severe enough. Benzothiophene, in solvent extraction, behaves very much like any other single ring aromatic, so some is removed with the extract. Clay treating, solvent dewaxing, and hydro-dewaxing have no apparent affect on benzothiophene.

The effects on polar nitrogen compounds (Figs. 23 and 24) can be illustrated with quinoline as an example. Acid treating will remove it, as will the subsequent clay treating. Hydrotreating will convert a small amount, and hydrogenation can, if desired, convert all of it to a saturated hydrocarbon. In solvent extraction, quinoline behaves like an aromatic, so it is partitioned between the raffinate and extract phases, with most going to the extract phase. The dewaxing processes would not affect the quinoline content.

Acid Treating

$$R\text{-}\underset{H}{C}\text{=}CH_2 + H_2SO_4 \longrightarrow R\text{-}\underset{\underset{CH_3}{H}}{C}\text{-}OSO_3H$$

(Alkyl Hydrogen Sulfate)

Removed With Sludge Or In Neutralization Step

Clay Treating – Limited Adsorption

Hydrotreating/Hydrogenation

$$R\text{-}\underset{H}{C}\text{=}CH_2 + H_2 \xrightarrow{\text{Cat.}} R\text{-}CH_2\text{-}CH_3$$

Saturated Hydrocarbon

Solvent Extraction
Solvent Dewaxing   No Effect
Hydro-Dewaxing

FIG. 18—*Hydrocarbon reactions: olefins.*

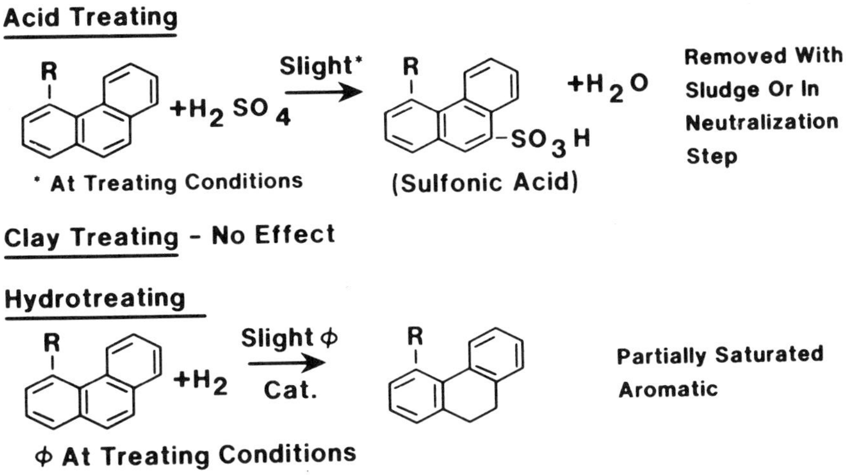

Acid Treating

R + H₂SO₄ → (Slight*) R—SO₃H + H₂O   Removed With Sludge Or In Neutralization Step

* At Treating Conditions   (Sulfonic Acid)

Clay Treating - No Effect

Hydrotreating

R + H₂ → (Slight φ / Cat.) R   Partially Saturated Aromatic

φ At Treating Conditions

FIG. 19—*Hydrocarbon reactions: aromatics.*

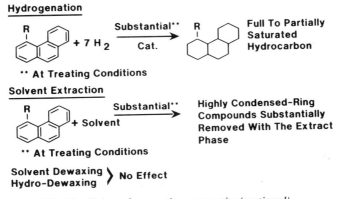

Hydrogenation

R + 7 H₂ → (Substantial** / Cat.) R   Full To Partially Saturated Hydrocarbon

** At Treating Conditions

Solvent Extraction

R + Solvent → (Substantial**)   Highly Condensed-Ring Compounds Substantially Removed With The Extract Phase

** At Treating Conditions

Solvent Dewaxing ⟩ No Effect
Hydro-Dewaxing

FIG. 20—*Hydrocarbon reactions: aromatics (continued).*

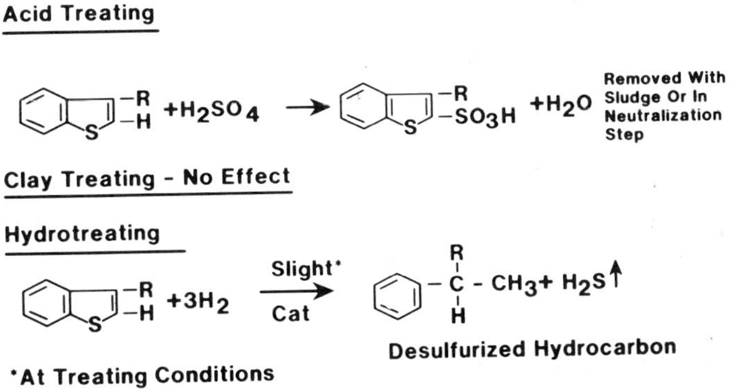

Acid Treating

—R / —H + H₂SO₄ → —R / —SO₃H + H₂O   Removed With Sludge Or In Neutralization Step

Clay Treating - No Effect

Hydrotreating

—R / —H + 3H₂ → (Slight* / Cat) ⬡—C(R)(H)—CH₃ + H₂S↑

Desulfurized Hydrocarbon

*At Treating Conditions

FIG. 21—*Hydrocarbon reactions: benzothiophenes.*

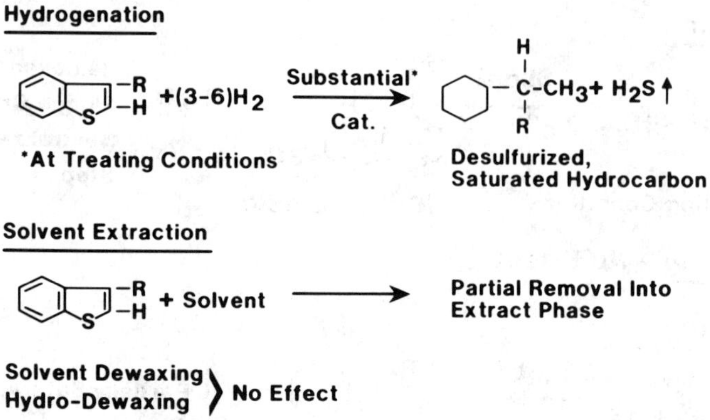
Hydrogenation
+(3-6)H2
Substantial*
Cat.
C-CH3+ H2S
*At Treating Conditions
Desulfurized,
Saturated Hydrocarbon
Solvent Extraction
+ Solvent
Partial Removal Into
Extract Phase
Solvent Dewaxing
Hydro-Dewaxing
No Effect

FIG. 22—*Hydrocarbon reactions: benzothiophenes (continued).*

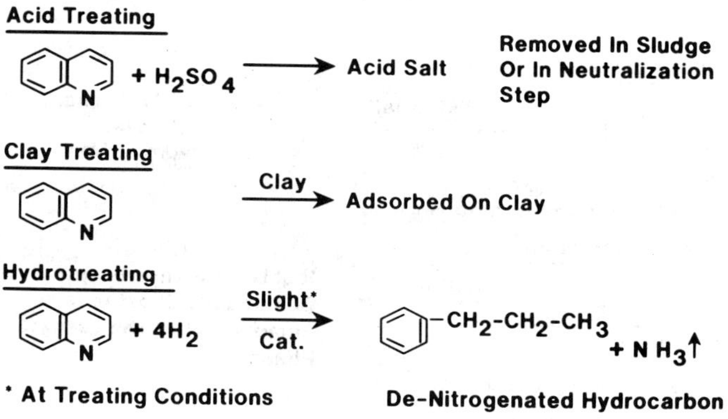
Acid Treating
+ H2SO4
Acid Salt
Removed In Sludge
Or In Neutralization
Step
Clay Treating
Clay
Adsorbed On Clay
Hydrotreating
+ 4H2
Slight*
Cat.
-CH2-CH2-CH3
+ N H3
* At Treating Conditions
De-Nitrogenated Hydrocarbon

FIG. 23—*Hydrocarbon reactions: polar nitrogen compounds.*

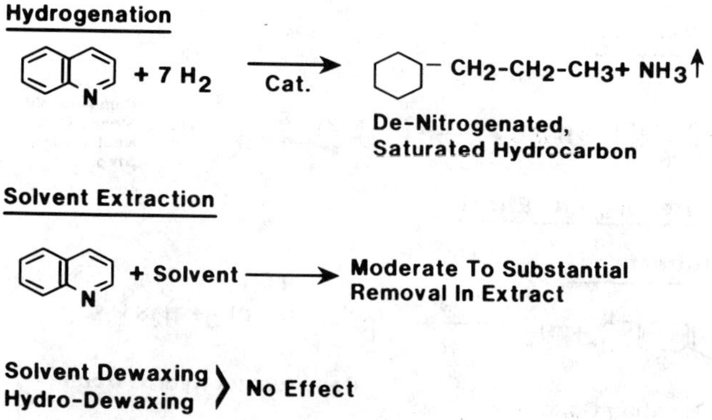
Hydrogenation
+ 7 H2
Cat.
CH2-CH2-CH3+ NH3
De-Nitrogenated,
Saturated Hydrocarbon
Solvent Extraction
+ Solvent
Moderate To Substantial
Removal In Extract
Solvent Dewaxing
Hydro-Dewaxing
No Effect

FIG. 24—*Hydrocarbon reactions: polar nitrogen compounds.*

Acid Treating

(1) $2RSH + H_2SO_4 \longrightarrow R_2S_2 + 2H_2O + SO_2 \uparrow$

(2) $R_2S_2 + H_2SO_4 \longrightarrow 2\ RS \cdot SO_3H \longrightarrow$ Removed With
Sludge Or In
Sulfonic Acid   Neutralization Step

Hydrotreating/Hydrogenation

$R-SH + 2H_2 \xrightarrow[\text{Cat.}]{} RH_3 + H_2S \uparrow$
Desulfurized Hydrocarbon

$R_2S_2 + 5H_2 \xrightarrow[\text{Cat.}]{} 2\ RH_3 + 2H_2S \uparrow$
Desulfurized Hydrocarbon

Clay Treating
Solvent Extraction $\longrightarrow$ No Removal
Solvent Dewaxing
Hydro-Dewaxing

FIG. 25—*Hydrocarbon reactions: polar sulfur-mercaptans and disulfides.*

Polar sulfur compounds, illustrated by mercaptans and disulfides in Fig. 25, are removed in acid treating and converted easily by hydrotreating and hydrogenation. The other processes do not seem to affect these compound types.

Polar oxygen compounds, illustrated by naphthenic acids in Fig. 26, are removed in acid treating and converted easily by hydrotreating and hydrogenation. Clay treating also is effective in removing naphthenic acids, hence its prevalent use in reclaiming used transformer oil. The other processes do not affect naphthenic acids.

Water and particulates are removed by deliberate drying and filtering, but can also be removed incidentally in some of the processes. Unfortunately, clay beds, catalyst beds, etc. make excellent filters. Over a long period of time, this may produce flow problems, such as pressure drop, in such operations.

Now that the refining processes and their chemical effects have been examined in some detail, let us turn to a different way of listing the effects that corresponds to our normal perception of a transformer oil. Table 2 illustrates the effects on the bulk oil properties that are of major interest. Most of the effects are self-evident from the previous discussions, but a few need comments.

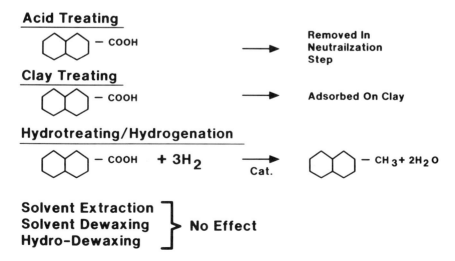

FIG. 26—*Hydrocarbon reactions: polar oxygen compounds-naphthenic acid.*

TABLE 2—*Unit process effects on oil properties.*

Quality	Acid Treat	Clay Treat	Hydro-treat	Hydro-genation	Solvent Extraction	Dewax	Pour Depress	Dry
Color	B	B	B	B	B			B
Color stability	B	B	B	B	B			
Pour point				S	S	B	B	
Flash point								
Interfacial tension	B	B	B	B	B			
Specific gravity	D		D	D	D	I		
Viscosity	D		D	D	D	I		
Aniline point	I		I	I	I	D		
Oxidation stability	B	B	B	W	B			
Power factor	B	B	B	B	B			B
Dielectric strength	B	B						B
Impulse strength	B		B	B	B			B
Gassing tendency	W		W	W	W			

NOTE: B = better; W = worse; I = increase; D = decrease; S = slight.

Hydrogenation as considered in this paper will, by itself, make the oxidation stability worse. Combined with the use of added oxidation inhibitors, it makes the oxidation stability much better.

The flash point is largely unaffected by any of the processes except distillation. The degree of stripping in some of the processes can marginally affect the value.

The gassing tendency is made poorer by any process that removes aromatics. Since none of the refining processes discussed increase aromatics, there is no way described herein to improve this property. The need to preserve a good gassing tendency must be balanced against the desire for a good impulse strength.

As mentioned earlier, it is possible to make a satisfactory transformer oil by many different process sequences. Some of the various combinations that have been reported are:

1. Acid treating, clay treating.
2. Hydrotreating, clay treating.
3. Hydrogenation.
4. Solvent extraction, hydrotreating.
5. Solvent extraction, hydrotreating, clay treating.
6. Solvent extraction, hydrotreating, dewaxing.
7. Hydrotreating, solvent extraction, dewaxing, clay treating.

These are probably not the only refining sequences used, and there will be others used in the future as the refiners cope with changing crude oils and new technology.

Table 3 shows a limited comparison between transformer oils made from differing crude sources and types using different processing. This table illustrates the fact that it should be possible to make a satisfactory conventional transformer oil from any good lube crude though paraffinic crudes require at least one additional processing step.

OSHA Regulation Effects

Adding complexity in today's environment in the United States is the Occupational Safety and Health Administration's (OSHA) Hazard Communication Standard, 29 CFR 1910.1200, which became effective 25 Nov. 1985. This regulation was subsequently modified in a Notice of Interpretation published in the *Federal Register* on 20 Dec. 1985 at 50 FR 51852. Under the requirements of this standard, some mineral transformer oils are classified as potentially carcinogenic based only on processing itself.

A full discussion of the OSHA standard and the Notice of Interpretation would require a separate presentation, but the results can be summarized as follows:

1. Oils requiring a cancer hazard warning:

   a. Oils made by acid treating.
   b. Oils made by mild hydrotreating.
   c. Oils made by mild solvent refining.

2. Oils exempt from a cancer hazard warning:

   a. Oils made by severe solvent refining.
   b. Oils made by severe hydrotreating.
   c. Oils made by sequential processing of mild hydrotreating and mild solvent refining.

OSHA has defined mild hydrotreating, but has not yet defined either mild or severe solvent refining.

Under the OSHA standard, oils made by acid treating plus clay treating and oils made by mild hydrotreating plus clay treating now require a cancer hazard warning. There has been a shift away from these two processes by the industry.

TABLE 3—Comparison of transformer oil properties.

Oil	A	B	C	D	E	F
CRUDE TYPE:	N	N	N	N	P	P
CRUDE SOURCE:	USA-1	USA-1	USA-2	Foreign	Foreign	Foreign
PROCESSING:[a]	AT, CT	HT, CT	SE, HT	SE, HT	SE, HT, DW	SE, HT, DW
PROPERTIES						
Specific gravity, 60/60°F	0.8855	0.8860	0.8735	0.8666	0.8624	0.8423
Color, ASTM	L0.5	L0.5	L0.5	L0.5	L1.0	0.5
Flash point, °C	152	154	154	148	191	158
Pour point, °C	−56	−56	−56	−51	−48	−45
Viscosity at 40°C, cSt	9.4	9.3	8.7	8.9	11.4	7.4
Neutralization no., mg KOH/g	Nil	Nil	0.01	0.01	Nil	0.02
Interfacial tension, dynes/cm	50.4	49.4	49	45	46	N/A
Sulfur, mass%	0.06	0.06	0.14	0.25	0.5	N/A
Aniline point, °C	74	74	78	80	85	89
Clay/silica gel analysis						
% Saturates	73.1	73.1	76.2	78	67.5	N/A
% Aromatics	26.7	26.7	23.6	22	32.1	N/A
% Polars	0.2	0.2	0.2	...	0.4	N/A
Power factor at 100°C, %	0.30	0.14	0.05	0.25	0.16	0.25
Dielectric breakdown voltage, kV	35	35	44	52	35	30
Impulse breakdown voltage, −kV	155	173	174	166	196	155
Gassing tendency, μL/min.	+4	−3	−6	−28	−10	+3
Oxidation stability, D 2440						
At 164 h, % Sludge	N/A	0.10	0.13	0.08	0.003	0.10
Acidity, mg KOH/g	N/A	0.28	0.35	0.28	0.03	0.12

[a]AT = acid treating; CT = clay treating; HT = hydrotreating; SE = solvent extraction; DW = dewaxing.

For more detailed information on this subject the reader is referred to the OSHA standard published in the *Federal Register* on 25 Nov. 1983 at 48 FR 53280 and to the mandatory reference of most significance, IARC (International Agency for Research on Cancer) Monograph Volume 33 (Pages 90, 150–151).

Handling and Storage

Only a brief discussion of handling and storage is included here. This is a subject worthy of a separate discussion also.

Transformer oil is a delicate product that must be treated with extreme care and dedication to prevent contamination. Water and particulates are the most common contaminants and the easiest to remove. There are other materials, such as motor oils that are extremely difficult to remove and must particularly be avoided.

Transformer oil received in drums, pails, etc. is subject to water contamination during storage. The bung closures and seals are not perfect and may over a period of time begin to "breathe". The mechanism is simple. The container becomes heated during the day and cools at night. The resultant expansion and contraction can be enough to "pump" moist air in and out of a drum sitting upright on a slab. This is accentuated, if the drum is stored outside.

It is preferred that drums be stored indoors and in racks laying on their sides with the two bungs horizontally opposed so as to keep the closures under the surface of the oil. This will greatly minimize chances of "breathing". Drums stored outdoors should never be stored upright. Water will collect on the head and can be pulled into the drum if "breathing" occurs.

Finally, keep the drum supply "fresh". Do not expect drums two and three years old to contain dry oil. They will not.

Transformer oil transported in bulk is subject to other sources of contamination. It should be transported in tank trucks or tank cars that are dedicated to that service alone. It is also desirable to have the conveyances equipped with dessicant breathers or nitrogen blanketing to reduce moisture contamination. Dessicant breathers may not be needed on tank trucks, if the delivery time is relatively short.

Tank trucks or tank cars that have been in service handling compounded oils (motor oils, turbine oils, etc.) or any vegetable or animal oil should not be used to haul electrical oils. The dissipation factor of the electrical oil will be adversely affected by these products. Conveyances that have been in these services are almost impossible to clean adequately for electrical oil use.

It is strongly recommended that any system for receiving transformer oil in bulk do so with totally dedicated hoses, lines, pumps, and tanks. It is also advisable to have the storage tank equipped with dried air or dried nitrogen blanketing and to provide a system to redry and filter the oil.

Drying is best accomplished with a vacuum dehydration unit or by means of dried gas (air or nitrogen) sparging in the tank. If sparging is used, a continuous purge of the gas space above the oil with dried gas will be required to prevent recondensation of the moisture on the tank roof.

It may seem strange to have a refiner caution the receiver of bulk product by saying the following: "Check the quality of the transformer oil, before it is unloaded." Nevertheless, there is justification to *always* check the quality of the oil upon receipt. It will save time, work, and anxiety in the long run.

This discussion of transformer oil handling has been brief. It, too, is a subject worthy of a separate discussion.

Conclusion

This paper has presented an overview of the manufacture of mineral insulating oils of petroleum origin and related topics. No startling new concepts have been presented, but rather it is intended that this discussion will improve the reader's understanding and appreciation of some of the complexities of transformer oil manufacture and handling.

Acknowledgment

The author wishes to express his indebtedness to the pioneering work of J. R. Lawley and Homer Jennings in comparing hydrotreating and acid treating as processes for refining mineral transformer oil.

DISCUSSION

C. L. Sobral Vieira[1] *(written discussion)*—According to ASTM D 3487, Type I oil can have up to 0.08% of DBPC. Is the permission given to the producer of the oil to add artificial inhibitor at this level a help to them to produce an oil that meets the specification (good oil) or is it a way to improve the life of an already good oil (an oil that would meet the specification even without the DBPC)?

T. Lipscomb (author's closure)—Both situations probably exist today. Remember that OSHA has said that a cancer hazard warning is required if the oil is made by mild hydrotreating plus clay treating or by acid treating plus clay treating. Some refiners have switched to severe hydrotreating to avoid a cancer hazard warning. Severe hydrotreating usually destroys enough of the natural inhibitors that addition of a synthetic inhibitor is required to pass the oxidation tests. Other methods that are severe but more selective may not require synthetic inhibitors. An example of this is solvent extraction plus hydrotreating. In this case the refiner may elect to add synthetic inhibitor to enhance the quality of an oil that already meets the oxidation tests.

[1]CEPEL, P.O. Box 2754, Rio de Janeiro, Brazil.

Howard C. Manger[1]

PCBs: The Beginning, the Muddle, the End?

REFERENCE: Manger, H. C., **"PCBs: The Beginning, the Muddle, the End?"** *Electrical Insulating Oils, STP 998,* H. G. Erdman, Ed., American Society for Testing and Materials, Philadelphia, 1988, pp. 25–34.

ABSTRACT: Over the years, there have been many changes in the federal regulations regarding PCBs. From the beginning, utility personnel were in a dilemma as to how to deal with the problem.

First, we had trouble realizing that the askarel we were using in our transformers was more than 50% polychlorinated biphenyl, and even if it was, could it possibly be a toxic or a hazard to anyone's health? After all, we and those before us had been up to our elbows in the stuff for years. It may have burned our eyes a little, but as far as we know it did us no physical harm. We also found it was many things in our everyday life, at work and at home.

Soon those of use close to the situation realized that the utility industry, in particular, was in deep trouble. Congress passed the Toxic Substance Control Act in 1976 and singled our PCB to be controlled. Three years later EPA published the first of a number of rules that have changed the way utilities do business. The industry has been forced to spend millions of dollars and countless man-hours dealing with regulations that we questioned from the start and, at best, learned to live with. In addition, the rules change so often it is truly difficult to keep up with them.

This paper tries to take us from the start, thru the confusing muddle of regulations, to where we are today.

KEY WORDS: askarel, Aroclor, Environmental Protection Agency, federal regulations, insulating fluids, polychlorinated biphenyls (PCBs), PCB, Toxic Substance Control Act

It was obvious in the early 1970s that Congress was serious about enacting legislation regarding control of toxic substances in the environment and in commerce. Polychlorinated biphenyls apparently were *not* one of the major concerns early in the process.

However, during the Senate debate on S.3149, March 1976, Senator Nelson introduced an amendment on polychlorinated biphenyls. His remarks were primarily on the control of PCB manufacturing and the environmental and health effects from PCB environmental contamination.

In the House of Representatives in August 1976, Representatives Dingell and Gude introduced an amendment on polychlorinated biphenyls. Their remarks were primarily on environmental levels and exposure. Other remarks by members were addressed to the question of substitutes for commercial PCBs and to why PCBs should be singled out and not controlled by other sections of the Act.

The Toxic Substance Control Act (Public Law 94-469) was enacted on 11 October 1976. Section 6(e) of the act required that polychlorinated biphenyls specifically be regulated. It is the only substance that was singled out.

In 1976, the cumulative U.S. production of PCBs since 1919 was estimated at 1.4 billion pounds. Of the 1.25 billion pounds used in the United States, about 760 million pounds were in service; 290 million pounds are in dumps and landfills; 55 million pounds had been destroyed; and 150 million pounds were believed currently located in soil, water, air, and sediment.

At that time, there were continuous discussions within EPA on the definition of "Polychlori-

[1]General supervisor, Baltimore Gas and Electric, Baltimore, MD 21203.

nated Biphenyls" as a chemical and on the scope of coverage which should be included by the definition. Section 6(e) only mentions "Polychlorinated Biphenyls" and does not use any other commonly used acronym, trade name, or chemical name; nor is there offered a definition.

It was clear to EPA from the wording of the Act and from the debates on the amendments that Congress meant for the Agency to control manufacturing, processing, distribution, use, and disposal of polychlorinated biphenyls, and the environmental contamination from all sources. Therefore, the marking and disposal regulations were to apply not only to newly manufactured PCBs, but also to those PCBs in use in electrical and other industrial equipment and those PCBs which occur as a significant impurity in other chemicals.

Commercial PCB

Commercial polychlorinated biphenyls or PCBs are prepared with anhydrous chlorine reacting with biphenyl in the presence of ferric chloride. This crude PCB mixture is distilled to obtain the finished product.

The biphenyl (diphenyl, phenylbenzene) ring is numbered as shown in Fig. 1.

Chlorine can be placed at any or all of the sites. There are 209 possible chlorobiphenyl isomers, but only about 100 different isomers have been found in commercial PCBs.

The synthesis of chlorinated biphenyl was reported as long ago as 1881, and successful commercial production of PCBs was commenced in this country in 1929 by the Monsanto Co., which has been the sole U.S. producer of Aroclors (Monsanto trade name).

The molecular composition of six typical Aroclors is shown in Table 1. Monsanto adopted a four-digit designation for its Aroclors; the last two digits indicate the approximate chlorine content by percentage weight; the first two digits indicate the type of material: biphenyl, triphenyl, or a mixture of the two.

Physical Properties

The three most important physical properties of PCBs are: low vapor pressures, low solubility, and high dielectric constants. They are miscible with most organic solvents.

The chemical properties that make PCBs desirable industrial materials are their excellent thermal stability, their strong resistance to both acid and basic hydrolysis, and their general inertness. They are quite resistant to oxidation.

The higher chlorinated biphenyls are nonflammable and have extremely low volatilities. Unfortunately, some of the same characteristics, such as stability and nondegradability, make them highly persistent in the environment.

Uses

Their chemical stability, low volatility, high dielectric constant, and compatibility with other chlorinated hydrocarbons have resulted in many and varied industrial applications for the PCBs.

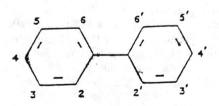

FIG. 1—*Biphenyl ring.*

TABLE 1—*Molecular composition of typical Aroclor.*

	Aroclor					
	1221	1016	1242	1248	1254	1260
Biphenyl	11	T	T	T	T	. . .
Monochloro	51	1	1	T	T	. . .
Dichloro	32	20	16	2	05	. . .
Trichloro	4	57	49	20	1	T
Tetrachloro	2	21	25	40	21	T
Pentachloro	T	1	8	32	48	12
Helachloro	. . .	T	1	4	23	28
Heptachloro	. . .	. . .	T	2	6	41
Octachloro	. . .	. . .	. . .	T	T	18
Nonachloro	. . .	. . .	. . .	. . .	. . .	1

NOTE: T = trace.

Major uses of PCBs were as dielectric liquids in capacitors (99% PCB) and in transformers (60 to 99% PCBs), as dye carriers in carbonless copy paper (3¹/₂% PCB in the paper), and as plasticizers in various paints and plastics (1 to 10% PCB).

Uses of PCB also included formulation into ballasts for fluorescent fixtures; impregnation for insulation of electric cables and electronic capacitors; and plasticizers of vinyl chloride polymer films.

Because of their thermal stability and fire resistance, PCBs also found application in high-pressure hydraulic fluids, specialized lubricants and gaskets, heat transfer agents, and machine tool cutting oils. Miscellaneous uses include: formation into some epoxy paints; protective coating for wood, metal, and concrete; adhesives; and in carbonless copy paper.

Askarels

Dielectric and cooling fluids used in electric equipment that contain significant amounts of PCBs and blends of chlorinated benzenes are generally referred to as askarels.

Manufacturers either modified the Aroclor themselves or had Monsanto prepare special formulations for them. These "askarel" have been marketed and used by the manufacturers in their products under various trade names. The more common names are listed in Table 2.

Toxic Substance Control Act (TSCA) (Start of the Muddle)

Congress passed TSCA in 1976 and appointed EPA the watchdog. Since then, we have been in a continuing muddle as how to deal with PCBs. One must realize that, with the swipe of a pen, Congress legislated a fluid used with no obvious ill effects into a hazardous substance.

However, the Toxic Substance Control Act allowed EPA to write regulations so there could be three ways for continued use of a PCB:

1. Totally enclosed.
2. Use authorization.
3. Exemption.

Totally Enclosed

The main benefit of having a use considered totally enclosed is that it may fall outside of the Toxic Substance Control Act if it can be shown that it offers insignificant exposure to human

TABLE 2—*Various trade names for askarel.*

Name	Manufacturer
COMMON DOMESTIC TRADE NAMES	
Aroclor	Monsanto
Aroclor B	Mallory
Asbestol	American Corp.
Askarel	Hevi-Duty Corp.
Askarel	Ferranti-Packard Ltd.
Askarel	Universal Mfg. Co.
Chlorextol	Allis-Chalmers
Clorinol	Spraque Electric
Clorphen	JARD Company, Inc.
Diaclor	Sangamo Electric
Dykanol	Cornell Dubilier
Elemex	McGraw Edison
Eucar	Electric Utilities Corp.
Hyv	Aerovox
Inerteen	Westinghouse Electric
No-Flamol	Wagner Electric
Pyranol	General Electric
Saf-T-Kuhl	Kuhlman Electric
COMMON FOREIGN TRADE NAMES	
Clophen	Bayer (Germany)
Fenclor	Caffaro (Italy)
Kennechlor	Kenneclor (Japan)
Phenoclor	Prodelec (France)
DK	Caffaro (Italy)
Pyralene	Prodelec (France)
Solvol	USSR

health and the environment. Unfortunately, insignificant exposure, by EPA definition, is *no exposure.*

Use Authorization

However, EPA can authorize a use, not considered totally enclosed, if it can be shown that such use does not offer unreasonable risk. The Toxic Substance Control Act may allow this authorized use for extended periods of time. To reiterate: Subject to EPA approval, a use considered totally enclosed does not have to be regulated; however, if the risk is reasonable, a use can be authorized.

Exemption

Exemptions are applied for by the user, who must show no unreasonable risk, indicate effort for improvement, and reapply for the exemption every year.

Regulation

With these ground rules, the first rule was published on 31 May 1979. EPA looked at the use activities of PCB and deemed the use of capacitors and transformers to be totally enclosed. EPA felt that this equipment posed an insignificant exposure to the environment. EPA also authorized a number of uses, some of which were railroad transformers, small capacitors, and carbon paper. There were a few exemptions sought under that 31 May 1979 final rule.

The rule established regulatory cutoff of 50 ppm for manufacturing, processing, distribution in commerce, or use of PCB.

It covered:

1. Marking.
2. Recordkeeping.
3. Restrictions on repairs.
4. Disposal (basically, incineration was the recommended method).
5. Storage.
6. Inspection.

This May 1979 rule took effect on 2 July 1979.

Rule Challenged

As new and disruptive as the regulation was, electrical equipment owners were learning to live with it. However, the Environment Defense Fund (EDF) challenged the 31 May 1979 rule in the U.S. Court of Appeals in Washington D.C. on three grounds:

1. EDF felt that there was no record for EPA to establish a 50-ppm cutoff.
2. That EPA had no data to back up the claim that capacitors and transformers were totally enclosed.
3. The authorizations that were allowed were improper.

The court ruled that transformers and capacitors were not totally enclosed, and that EPA lacked substantial evidence to support a regulatory cutoff of 50 ppm for manufacture, processing, distribution in commerce, or use of PCB.

Had the court's decision gone into effect, making any use of PCB illegal, EPA said it would have caused a major economic impact. I think we would have had to turn the lights out.

However, EPA, EDF, and certain industry and electric utility intervenors involved in the case filed a joint motion seeking a stay of the court's mandate until further rulemaking could be completed. The court granted this request.

Additional Regulation (as a Result of the Court Decision)

On 22 April 1982, EPA issued a proposed rule governing the use and servicing of electrical equipment containing PCBs. The final rule appeared in the *Federal Register* of 25 August 1982. This final rule was issued as a result of the court's decision and dealt only with electrical equipment. It authorized certain uses of PCB in electrical equipment and set 50 ppm as a cutoff for electrical equipment.

This Rule:

1. Prohibited the use of PCB transformers and PCB electromagnets posing a risk to food and feed facilities after 1 Oct. 1985.

2. Expanded the definition of electrical equipment posing an exposure risk to food or feed facilities.

3. Authorized the use of all other PCB transformers for the remainder of their useful lives, with inspections. (Take note: it "authorizes," where May 1979 considered transformers totally enclosed.)

4. Set requirements for inspection frequency and retention and availability of records.

5. Authorized the use of all PCB-containing mineral oil-fill electrical equipment.

6. Presumed circuit breakers, reclosers, and cables to contain less than 50-ppm PCBs.

7. Authorized the use of large PCB capacitors located in electrical substations and indoor installations.

8. Prohibited the use of all other large PCB capacitors after 1 Oct. 1988. (Here we see a time period tied to authorization.)

9. Allowed storage of large PCB capacitors and PCB-contaminated equipment outside of qualified storage facilities.

10. Defined "disposal" as including leaks and spills, but went on to say: spills, leaks, or uncontrolled discharges resulting from the authorized use (or storage) of electrical equipment shall not constitute a disposal violation provided adequate cleanup measures are initiated within 48 h after notice of the discharge.

11. Required records of inspection and history to be kept for three years after disposing of PCB transformers.

Nonelectrical Rules

The 25 Aug. 1982 rule authorized a 50-ppm cut only for electrical equipment. What about the other items and systems that have PCB in them (even a molecule of PCB)? There are many nonelectrical items that fall into this category.

EPA has stated that any time you have a carbon, chlorine, and some heat, the possibility exists that you can create a polychlorinated biphenyl. It follows that any manufacturing process that uses these parameters is suspect of the incidental generation of PCBs. Industrial representatives convinced EPA that there are some manufacturing systems in which PCB may be created, but where it is not released into the environment. EPA issued Phase One of a two-part rule on 21 Oct. 1982 to address the 50-ppm regulatory cutoff in "closed and controlled" waste manufacturing processes.

The second phase of regulation dealt with processes involving "uncontrolled" PCBs generated in other than "closed" and "controlled waste" processes.

These non-Aroclor, inadvertently generated PCBs were the principal subject of a 10 July 1984 rule. The exclusions announced in this 1984 rule expanded upon and superseded the exclusions for closed and controlled waste manufacturing processes. The generic exclusion for inadvertently generated PCBs applied to manufacturing processes which qualified as "excluded manufacturing processes." These excluded processes were defined in terms of established limits for PCB releases in products, air emissions, water effluents, and wastes.

Transformer Fires and the Regulation

At the time these rules were written, information available to the EPA indicated that fires involving electrical transformers were rare, isolated incidents. However, several transformer fires in buildings brought into question EPA's earlier assumptions.

Binghamton, NY

The first and most notable event involved an 18-story state office building, the tallest and most prominent landmark in Binghamton, a city of some 60,000 people in southeastern New York. On 5 Feb. 1981, an intense fire occurred in the electrical switching equipment in the basement mechanical room of the building. The switching equipment was located adjacent to two transformers filled with askarel (65% polychlorinated biphenyls and 35% chlorinated benzenes). The building and transformers were owned by the State of New York.

The heat from some allegedly burning automobile fan belts stored next to the transformers cracked a porcelain bushing on one of the transformers and about 180 gal of askarel sprayed onto the floor and into the hot switch gear. The fire produced a fine soot that spread up an open vertical chase which carried the exhaust throughout every floor of the building.

Immediately following the fire, and before the extent and nature of the contamination became known, more than 300 people reentered the building to help with the cleanup or to retrieve documents and personal effects. The cleanup effort was begun. However, three weeks after the fire, the extent of the contamination was discovered.

Analysis of soot samples recovered from the building indicated high concentrations of PCBs and the presence of polychlorinated dibenzodioxins at about 3 ppm and dibenzofurans at concentrations of around 300 ppm.

The building was closed and secured and cleanup was started. Every movable item was scrapped. Walls and partitions were ripped out and thousands of barrels of material were hauled to the Niagara Falls landfills, including six acres of floor tiles. The State had hoped to reopen the building in April 1985. It has not opened yet (January 1988) and the cost of cleanup, so far, is approaching twice the original cost of the building.

San Francisco

About two years later, on 15 May 1983, a fire occurred in a transformer vault located under a sidewalk adjacent to One Market Plaza, an office building in downtown San Francisco. The transformers involved in Binghamton were inside the building and were the property of the owner. This askarel transformer was outside in a sidewalk vault and was owned by Pacific Gas & Electric Co. The smoke from this incident was drawn into the building thru an outside intake vent adjacent to the sidewalk grating. The lower floors of this building were contaminated with PCB and furans. Although the upper floors were reoccupied after a short time, it took months before the lower floors were cleaned to a level low enough for reopening.

Tulsa

A similar sidewalk-type transformer vault incident occurred about a year later in Tulsa, Oklahoma. The high side switch of a network transformer overheated and exploded, atomizing between 20 and 30 gal of PCB onto a city street and into an adjacent basement of the Beacon building. A cleanup was commenced by Public Service of Oklahoma, who owned this transformer.

Subsequent sampling for PCB indicated that the cleanup was not successful, and a professional cleanup company with experience at the San Francisco fire was hired.

For the protection of their workers, this contractor tested for the presence of furans and dioxins. These tests showed the presence of furans in surprisingly large quantities. The basement was again evacuated some four months after the accident, and a complete cleanup was commenced.

The cleanup of the basement was in its very last stages when a second identical explosion occurred in the high side switch of the replacement transformer. This time there was considerable contamination on the outside of the building and on the outside of an adjacent building containing a hot dog restaurant and bar, but, because of the work performed in the vault, no PCBs penetrated the basement/vault wall of the Beacon building.

However, before a cleanup of the vault could commence, a 12-in. rain deluged Tulsa and flooded the vault. This water breached the basement wall of the Beacon building with PCB-contaminated water, requiring another involved and costly cleanup.

Transformer Fire Rule

Because of these and other occurrences, the Agency issued an advance notice of proposed rulemaking (ANPR) on 23 March 1984, a proposed rule on 11 Oct. 1984 to gather data on the specific risks posed by fires, and a final rule for electrical transformers on 17 July 1985.

Briefly, EPA imposed the following amendments to the August 1982 Rule. This rule required immediate notification to the National Response Center of a PCB transformer fire-related incident and prohibited the further installation of PCB transformers in or near commercial buildings. A number of requirements were to be completed by 1 Dec. 1985:

1. The registration of all PCB transformers with fire departments or fire brigades with primary response function.
2. The registration of all PCB transformers owned by utilities located in or near buildings, with the owners of the building.
3. The marking of the exterior of all PCB transformer locations (excluding grates and manhole covers).
4. The removal of combustible materials stored within 5 m of a PCB transformer enclosure.

The regulation also set another date as a milestone: 1 Oct. 1990. It prohibited the continued use of higher secondary voltage network PCB transformers (equal to or greater than 480 V) in or near commercial buildings beyond 1 Oct. 1990. This final rule also required by 1 Oct. 1990 the installation of enhanced electrical protection on lower secondary voltage *network PCB transformers* (less than 480 V) and on all radial PCB transformers used in or near commercial buildings.

Challenge to the Fire Rule

The 17 July 1985 rule, which became known as the "fire rule," was challenged by Mississippi Power Co. in the U.S. Court of Appeals for the 5th Circuit in Louisiana. In the settlement of that appeal EPA agreed to undertake good faith efforts to issue and cause to be published in the *Federal Register* by 31 Dec. 1986, a document clarifying EPA's position concerning the interpretation of certain provisions of the rules. (The Document is a series of questions and answers. EPA made the deadline. They were published on the last day, 31 Dec. 1986.)

In addition, some important salient amendments to the PCB Rule were proposed. The restriction in the "fire rule," that would not allow a PCB transformer to be installed in a location in or near a commercial building has been changed to allow such in (1) emergency situations (EPA states they would not want to "deny electric service to the entity or entities served by such transformer") and (2) "in situations where the transformer has been retrofilled and is placed in service in order to qualify to be reclassified."

Although the requirement that a network PCB higher voltage transformer, that is, with secondary greater than or equal to 480 V in or near commercial buildings, must be reclassified to less than 500 ppm or be replaced by 1 Oct. 1990 has not changed, the settlement amendment states that radial PCB transformers in or near commercial buildings must be equipped with electrical protection by 1 Oct. 1990, which can deenergize the transformer within several hundreds of a second. (The 17 July 1985 rule called for several tenths of a second.)

PCB network lower voltage transformers (less than 480 V) have been further reclassified into those located in sidewalk vaults and those not located in sidewalk vaults. Those located in sidewalk vaults must be removed from service by 1 Oct. 1993. The ruling is not specific, but I would think reclassification is an option.

PCB network lower voltage transformers not located in sidewalk vaults may continue in service after 1993 if electrical protection is supplied to deenergizing the transformer within several tenths of a second. This must be done by 1 Oct. 1990. In addition, if a transformer has not been so protected by 1 Oct. 1990, the transformer must be registered with EPA.

These changes to the "fire rule" should be published in the spring of 1988.

Spill Cleanup

A very important item has finally come to fruition. The administrator of EPA signed a PCB spill policy on 20 March 1987. It was published in the *Federal Register* as 761.20 on 2 April 1987.

There have been efforts to come to some sort of an agreement about spill cleanup with EPA since 1984. A number of organizations were involved, including the Utility Solid Waste Activities Group, EEI, NEMA, the Chemical Manufacturers' Association, the Environmental Defense Fund, and the Natural Resource Defense Fund. Early on, environmental factions and EPA were talking about cleaning up almost every molecule of spilled PCB. Further, for some time, EPA has considered a spill or a leak as improper or unauthorized disposal, and, as such, they could require a cleanup to background levels. It is a credit to all the organizations that this policy was the result of a consensus agreement.

There are a few salient points concerning this spill cleanup policy.

It is very significant that no cleanup requirements exist for leaks or spills of fluids containing less than 50 ppm PCB, that is, leaks or spills from non-PCB transformers are not regulated by this policy.

For any spill or leak of fluids containing more than 500 ppm (a PCB fluid), EPA requires not only cleanup of the fluid, contaminated soil, and contaminated surfaces, but also analytical sampling of the soil and surfaces to assure specific low concentrations of PCB following completion of the cleanup. These levels are as low as 10 ppm, depending on where the spill is, and are as low as 10 μg per 100 cm^2, again depending on where you might have spilled this fluid.

For spills and leaks of less than 270 gal of fluid between 50 and 500 ppm PCB, EPA requires only cleanup of visible traces of fluids and double rinsing of surfaces, but does not require analytical sampling to confirm the adequacy of the cleanup. These cleanups must be completed within 48 h of the time you were aware of the spill. However, if some of this fluid should enter a residence, the cleanup levels could be as low as 10 μg per 100 cm^2. Any spill or leak of material containing more than 10 lb of PCB must be reported to EPA.

One of the more important things that came out of the consensus and this policy is that penalties for improper disposal can be avoided if there is a good faith compliance with the cleanup requirements. Basically, this is saying that if you make a good faith effort to clean up the spill, and it is realized later that you didn't fully comply with the regulations, you can be asked to clean up to a lower level, but you won't be fined.

Health Effects

PCBs can certainly be classified as an environmental contaminant. They are stable, do not biodegrade easily, and tend to bioaccumulate in organisms. As they move up in the food chain, they may increase in concentration. Laboratory studies on fish and animals have shown PCBs to cause chronic (long-term) toxic effects in some species even at low concentrations; however, PCBs are not acutely toxic or lethal. Numerous health effect studies have been made by some of the most prestigious health consultants in the country. In summary, they conclude:

1. PCBs have not been found to be implicated as human carcinogens.
2. PCBs have not been shown to affect adversely human reproduction or cause birth defects.
3. PCBs do not affect enzyme production in humans in a manner that leads to adverse health effects.
4. The only proven adverse human health effect resulting from PCBs is chloracne or skin irritation, a reversible health effect resulting in no long-term adverse health consequences.

The End ??????

Available evidence today indicates that some level of concern about PCBs is well founded. PCBs are persistent, ubiquitous, and bioconcentrate. Subsequent studies, however, show that adverse health effects have not resulted from past levels of exposure to PCBs. Future effects are even more unlikely because recent studies of PCB levels in the environment, in the food chain, and, indeed, in human adipose tissue are definitely trending downward.

It should then follow that additional PCB regulations are unnecessary, and it is time to move on to other issues. While this may be logical and reasonable based on past experience, it is doubtful that this is the end of the ever-expanding PCB regulations.

Bibliography

Environmental Research, Vol. 5, No. 5, September 1972.

Wood, D., National Conference On PCBs, 19-20 Nov. 1975.

Druley, R. M. and Ordway, G. L., The Toxic Substances Control Act, The Bureau of National Affairs, Inc., Washington, DC, 1977.

Draft Preamble to Proposed PCB Marking and Disposal Rule, Part 761, 4 Feb. 1977.

Federal Register, 31 May 1979, Vol. 44, No. 106, Supplementary Information and Rule.

EPA's Final PCB Ban Rule, Over 100 Questions & Answers to Help You Meet These Requirements, June 1980.

Manger, H. C., "Status of EPA Rules and Regulations," Doble Conference Minutes, 1981.

"PCB Regulations Status Report," Doble Conference Minutes, 1982.

Federal Register, 25 Aug. 1982, Vol. 47, No. 165, Supplementary Information and Amendments.

Craddock, J. H., "The History and Toxicology of PCBs," PCB Seminar, Baltimore, MD, March 1983.

EPA's Final PCB Ban Rule, Over 100 Questions & Answers to Help You Meet These Requirements, Revision 3, August 1983.

Manger, H. C., "Progress Report on Status of EPA's Latest Rules," Doble Conference Minutes, 1983.

"Update on PCB Rules and Regulations," Doble Conference Minutes, 1984.

Federal Register, 11 Oct. 1984, Vol. 49, No. 198, Supplementary Information and Amendments.

Manger, H. C., "Reclassification of PCB and PCB Contaminated Transformers," Doble Conference Minutes, 1985.

"Fires Involving PCB Transformers—Further Regulations," Doble Conference Minutes, 1986.

Federal Register, 17 July 1985, Vol. 50, No. 137, Supplementary Information and Amendments.

Code of Federal Regulations 40 Part 700 to End, revised as of 1 July 1985.

Kump, R. K., "EPA Update: Fire Rules," Doble Conference Minutes, 1986.

Association of Edison Illuminating Companies, Report of Committee on Power Distribution 1985-1986, PCB Transformer Regulation Implementation.

Manger, H. C., "Status of EPA Rules and Regulations," Doble Conference Minutes, 1987.

Federal Register, 2 April 1987, Vol. 47, No. 250, Supplementary and Amendments.

Section II—Significance of Application

Dennis L. Johnson[1]

Insulating Oil Qualification and Acceptance Tests from a User's Perspective

REFERENCE: Johnson, D. L., **"Insulating Oil Qualification and Acceptance Tests from a User's Perspective,"** *Electrical Insulating Oils, STP 998,* H. G. Erdman, Ed., American Society for Testing and Materials, Philadelphia, 1988, pp. 37–45.

ABSTRACT: To be acceptable, insulating oil must have several specific electrical, chemical, and physical attributes. Tests that demonstrate that the oil does indeed possess the desired characteristics are required. The tests specified by one utility are outlined in the paper with limits listed. Also described in the paper are insulating oil uses on a utility system. An attempt is made to relate oil use to each required test. The relative significance of each test is addressed.

KEY WORDS: insulating oil, technical specifications, acceptance tests, qualification tests

My occupational involvement with insulating oil is through my employment as a manager in a large federal utility, the Bonneville Power Administration (BPA). BPA is the largest of five power marketing agencies in the United States Department of Energy. The BPA service area covers Oregon, Washington, Idaho, Western Montana, and corners of California, Utah, and Wyoming. We have electrical equipment located near the Pacific Ocean, in the high deserts of Eastern Oregon and Washington, and in the arctic-like environs of the Northern Rockies and Grand Tetons. Our equipment faces all weather extremes.

BPA transmits bulk power to its 100 plus customers in the Northwest United States. The customers are the local public and private utilities and electroprocess industries. There are more than 14 000 miles of transmission lines connecting some 400 substations. Installed in those substations are approximately 1100 large, high-voltage power transformers and reactors and approximately 1600 oil-insulated, high-voltage power circuit breakers. The facilities also include numerous other related oil-filled equipment items such as station service and instrument transformers, and voltage regulators. We estimate that altogether we have at least 6 000 000 gal of insulating oil in energized equipment. Of course, it goes without saying that the integrity of each gallon, both as insulant and as a media for heat transfer, is of considerable significance to the reliable transmission of electrical energy to the distribution or end-use customer.

Procurement Practice

For our utility, oil performance is especially important because our equipment specifications encourage manufacturers to minimize the amount of oil needed. Further, we specify reduced basic insulation levels (BIL) in equipment. Thus, tremendous reliance is placed on the oil to withstand the rigors of thermal and electrical stress during service life. Our practice is to buy the oil we use separately from major equipment. This practice gives us better control and knowledge of what liquid goes into our electrical equipment. Equipment suppliers are prenotified of

[1]Manager, Substation Maintenance, Bonneville Power Administration, Vancouver, WA 98666.

the oil we intend to use, and, with their consent, our new-product warranty is therefore preserved.

We prepare equipment for service ourselves. We have excellent oil-handling apparatus and our crews have developed substantial expertise through experience. In all cases, our equipment preparation practices either match or exceed manufacturers' requirements. Through past practice and thorough documentation of equipment preparation activity, a reputation has been established that withstands manufacturers' scrutiny in the event of trouble during the warranty period.

By being large and by virtue of being an agency of the federal government, it seems our procurements are uniquely complicated. The procurement process is a function of federal procurement regulations intended to promote competition and result in high value for the dollars spent. Whether or not this goal is attained is a topic for other symposiums. The purchase specifications for insulating oil that I am familiar with include a thick layer of "boiler plate," which also is best left for other discussions. What we do want to focus on is the examination given to insulating oils available on the market and what we learn about quality of individual oils from that examination.

Technical Specifications

The specification requirements for insulating oil used in electrical equipment at BPA are that the oil shall be in accordance with Type II of ASTM Specification for Mineral Insulating Oil Used in Electrical Apparatus (D 3487). Tests, including submittal of test reports and data, assure that the oil being furnished does meet the sampling and testing requirements of D 3487. Oil suppliers are also requested to provide information about:

1. Supplementary design as listed in Appendices X1 and X2 of D 3487.
2. The generic refining process used.
3. The type of crude.
4. The type of oxidation inhibitor added.
5. Polychorinated biphenyl (PCB) concentration, equipment used for PCB tests, and accuracy and minimum detection range of PCB test equipment. ASTM Method for Analysis of Polychlorinated Biphenyls in Mineral Insulating Oils by Gas Chromatography (D 4059) is to be followed.

The concentration of PCBs in the oil shall not exceed 2 ppm when tested according to D 4059.

Qualification Testing

Vendors of insulating oil are required to demonstrate that the product they offer is of suitable quality. To qualify to supply oil to the government, a vendor must furnish representative samples taken in accordance with ASTM Method for Sampling Electrical Insulating Liquids (D 923) for a rigorous testing regimen. Samples for qualification testing are taken after final refining processes are complete but before loading for transportation. Once a vendor becomes qualified, that testing process is not repeated unless a significant product change occurs.

Acceptance Testing

Qualified vendors must provide further samples for acceptance testing for individual deliveries. Acceptance testing involves only tests needed to assure that batches of known products retain the expected characteristics.

The specifications require that samples of oil for acceptance testing be drawn by the vendor from his storage tanks. We draw samples from our dedicated oil transport trucks after they are loaded. (Use of our own oil tank trucks has eliminated frequent disconcerting instances of contamination that occur during shipping.) Both samples are brought to BPA's Materials Laboratory in Vancouver, Washington for acceptance testing. The supplier's samples are the "referee" samples. On occasion we have taken additional samples from the transport vehicle at our site if we suspected the oil had become contaminated during transit.

The oil samples provided are examined in detail. Individual tests referenced in D 3487 are listed in Table 1 with a brief comment on each following. After this review of the D 3487 tests, the paper will be concluded with remarks about additional oil characteristics of importance that should be investigated.

TABLE 1—*Qualification & acceptance tests and limits.*

ASTM Method	Limit	Qualifi-cation	Accep-tance
ELECTRICAL TESTS			
Dissipation Factor	0.05%, max at 25°C	x	x
(Power Factor) (D 924)	0.3%, max at 100°C	x	
Dielectric Breakdown (D 877) (disks)	30-kV, min	x	x
Dielectric Breakdown (D 1816)	28-kV, min at 0.040-in. gap		
(spheres)	56-kV, min at 0.080-in. gap	x	x
Gassing Tendency (D 2300)	+15 µL/min, max (Procedure A)	x	
	+30 µL/min, max (Procedure B)	x	
Impulse Breakdown (D 3300)	145 kV, min at 25°C	x	
CHEMICAL TESTS			
Water (D 1533)	35 ppm, max	x	x
Acidity (D 974)	0.03 mg KOH/g, max	x	x
Corrosive Sulfur (D 1275)	Noncorrosive	x	
DBPC Content (D 2668)	0.3% by mass, max	x	
DBPC Content (D 1473)	0.3% by mass, max	x	
Oxidation Stability (D 2112)	195 minutes, min	x	x
Oxidation Stability (D 2440)	72 h; 0.1% sludge, max / 0.3 mg KOH/g acid, max	x	
PHYSICAL TESTS			
IFT (D 971)	40 dynes/cm at 25°C, min	x	
IFT (D 2285)	40 dynes/cm at 25°C, min		x
Color (D 1500)	0.5, max	x	x
Visual Exam (D 1524)	Bright and clear	x	x
Pour Point (D 97)	−40°C, max	x	
Flash/Fire Point (D 92)	145°C, min	x	
Aniline Point (D 611)	63–84°C, max	x	
Viscosity (D 88)	3.0 cSt at 100°C, max / 12.0 cSt at 40°C, max / 76.0 cSt at 0°C, max	x	
Viscosity (D 445)			
Specific Gravity (D 1298)	0.91 at 15°C, max		x
DESIGN TESTS			
Coefficient of Thermal Expansion (D 1903)			
Thermal Conductivity (D 2717)			
Specific Heat (D 2766)			
ADDITIONAL TESTS			
Analysis of Gases Dissolved in Insulating Oil by Gas Chromatography (D 3612)			
Static Charging Tendency			
Analysis of PCBs in Insulating Liquids by Gas Chromatography (D 4059)			
Dielectric Strength of Oil in Motion			
Specific Resistance (Resistivity) of Electrical Insulating Liquids (D 1169)			

Electrical Tests

ASTM Test Method for A-C Loss Characteristics and Relative Permittivity (Dielectric Constant) of Electrical Insulating Liquids (D 924)

Limit: 0.05% max at 25°C, 0.3% max at 100°C

Commonly called the power factor test, this is a valuable screening test. If new oil has a power factor greater than 0.05%, it may include water, oxidation products, or other polar contaminants or combinations of these. Oil testing above 0.05% is not acceptable and should be looked at very closely. The 25°C test is always performed for acceptance, whereas the 100°C test is made for qualification of new oils and not normally repeated unless problems are suspected.

ASTM Test Method for Dielectric Breakdown Voltage of Insulating Liquids Using Disk Electrodes (D 877)

Limit: 30-kV, minimum

This test determines if an oil has adequate insulating strength and is used as a vendor qualification test. It is also routinely run as an acceptance test on samples of new oil as received. Low breakdown voltages usually mean that free moisture and conducting particles are present in the oil. Filtration will normally remove contaminants that cause low dielectric strength. This test will not detect any moisture in solution, which could amount to as much as 45 to 50 ppm for new oil at 25°C.

ASTM Test Method for Dielectric Breakdown Voltages of Insulating Oils of Petroleum Origin Using VDE Electrodes (D 1816)

Limit: 28-kV, minimum at 0.040-in. gap
56-kV, minimum at 0.080-in. gap

This sensitive dielectric strength test is used as both an acceptance and as a qualification test, but since the method is highly sensitive to moisture, new oil has to be first treated to dehydrate it and to remove gases. The VDE dielectric test is also routinely used to test oil as received. Failure to meet the minimum strength limit due to moisture or gas content will not cause rejection because the oil will be dried and degassed before it is used in energized equipment. For new oils before processing, typical results range from 40 to 60 kV with a 0.080-in. gap. (We use the 0.080-in. gap exclusively.)

ASTM Test Method for Gassing of Insulating Oils Under Electrical Stress and Ionization (Modified Pirelli Method) (D 2300)

Limit: +15 µL/min, max (Procedure A)
+30 µL/min, max (Procedure B)

New equipment designs that reduce materials and costs may result in greater electrical stress on the liquid insulation, causing gases to be evolved or absorbed. Oils with slightly negative gassing tendencies are preferred. However, price and availability may override decisions to avoid oils with high gassing tendencies. Gassing tendency is an important qualification test, but we do not perform it. When needed, the services of an external laboratory would be obtained.

Test Method for Dielectric Breakdown Voltage of Insulating Oils of Petroleum Origin Under Impulse Conditions (D 3300)

Limit: 145-kV, minimum, at 25°C.

This is a crucial qualification test. Impulse strength tests demonstrate the ability of an insulating oil to withstand the voltage stress caused by transients from lightning or switching. The test can be done in the BPA Laboratory if needed. Our primary concern is with impulse strength of total transformer systems, however. The ability to withstand transient voltages is very dependent on geometry, spacing, and polarity.

Chemical Tests

ASTM Test Method for Water in Insulating Liquids (Karl Fischer Method) (D 1533)

Limit: 35 ppm, max

This test is used for both vendor qualification and routine acceptance. If sample handling has been sloppy, high readings not indicative of the true moisture level in the oil can result. To avoid misleading results, sampling procedures require: (1) use of clean, dry nonpermeable-to-air containers; (2) precautions to preclude ingress of moisture (which in the rainy Pacific Northwest can be challenging); and (3) speed in shipment of samples to the laboratory. Temperature/pressure cycling can draw water into the sample. Low moisture content in mineral insulating oil is needed to minimize metal corrosion, maximize life of cellulosic insulation, and achieve adequate electrical strength and low dielectric loss characteristics. Water in oil levels greater than 10 ppm are considered to be high and cause for concern. Levels below 10 ppm can be achieved and maintained without extreme efforts.

ASTM Test Method for Neutralization Number by Color-Indicator Titration (D 974)

Limit: 0.03 mg KOH/g max

This is one of the tests we rely on to determine when to reclaim service-aged oil. We also apply it for both vendor qualification and delivery acceptance purposes. It is important to us to always start with a low acid content oil (less than 0.01 mg KOH/g) to help attain full amortized life expectancy from installed equipment and to protect the investment in capitalized facilities. In our opinion, the acceptable limit should be reduced from 0.03 mg KOH/g max to 0.025 mg max KOH/g for new oil.

ASTM Test Method for Corrosive Sulfur in Electrical Insulating Oils (D 1275)

Limit: Noncorrosive

This is still a necessary test method even though corrosive sulfur is not a problem with most modern oils. The apparatus for the test is in our laboratory and the test is run for qualification purposes. Free sulfur or sulfur compounds in oil can result in corrosion of copper in transformers and may help form acids.

ASTM Test Method for 2,6-Ditertiary-Butyl Para-Cresol (DBPC) and 2,6-Ditertiary-Butyl Phenol (DBP) in Electrical Insulating Oil by Infrared Absorption (D 2668)

Limit: 0.3% by mass, max

This is a good semiquantitative test used for vendor qualification. Inhibitor content in oil is important to slowing down oxidation aging rates and maintaining long service life of transformers. Our laboratory believes the alternative gas chromatograph method now being developed by ASTM for measuring inhibitor content will provide more reliable quantitation.

ASTM Test Method for 2,6 Ditertiary-Butyl Para-Cresol (DBPC) in Electrical Insulating Oils (Intent to Withdraw) (D 1473)

Limit: 0.3% by mass, max

Method D 1473 is our preferred method for measurement of inhibitor content because the infrared absorption method (D 2668) is sensitive to mixtures of 2,6 ditertiary-butyl phenol and 2,6 ditertiary-butyl para-cresol.

ASTM Test Method for Oxidation Stability of Inhibited Mineral Insulating Oil by Rotating Bomb (D 2112)

Limit: 195 minutes, minimum

New oils inhibited with DBPC are tested according to D 2112 for qualification. Oils with oxidation stability are essential to long service life of equipment. Oils that do not accept inhibitors are less desirable. The test also gives a clue about the consistency of production processes at various refineries, which makes it useful for acceptance purposes as well.

ASTM Test Method for Oxidation Stability of Mineral Insulating Oil (D 2440)

Limit: 72 h; 0.1% sludge max, 0.3 mg KOH/g acid max

This oxidation stability test is a useful qualification test; however, D 2440 is not currently in use at BPA. The method assesses the resistance of an oil to oxidation under accelerated aging conditions. The Doble Power Factor Valued Oxidation (PFVO) test gives information adequate to assess oxidation stability. The PFVO test characterizes the power factor of an oil specimen under accelerated aging conditions, measuring oil quality up to the point sludging starts and revealing the sludge-free life of the specimen.

Physical Tests

ASTM Test Method for Interfacial Tension of Oil Against Water by the Ring Method (D 971)

Limit: 40 dynes/cm at 25°C, minimum

This test, which uses a platinum ring, is used by our laboratory for qualification of vendors. An alternate method [ASTM Test Method for Interfacial Tension (IFT) of Electrical Insulating Oils of Petroleum Origin Against Water by the Drop Weight Method (D 2285)], which is more practical for field application as a diagnostic aid, is used by the laboratory for acceptance purposes. The platinum ring method is, of course, used to officially reject any oil batches. IFT values less than 40 dynes/cm indicate that unacceptable polar contamination is present, making this test a useful screening device for new oil exposed during transport to soaps, acids, paints, varnishes, or solvents. Oils with low IFT initially may result in accelerated aging of oil and solid insulation in transformers. The limit of acceptability could probably be raised to 45 dynes/cm.

ASTM Test Method for ASTM Color of Petroleum Products (ASTM Color Scale) (D 1500)

Limit: 0.5 max

Because of this test, used for both acceptance and qualification, some suppliers are eliminated. Slightly colored new oil does not indicate poor product quality. This test is useful, however, in detecting deterioration during service (generally over long time periods). New oils that are bright and clear are desired. Clear oil (low color numbers) allows visual inspection of internal equipment components and is some assurance that the oil is probably in serviceable condition.

ASTM Method for Visual Examination of Used Electrical Insulating Oils of Petroleum Origin in the Field (D 1524)

Limit: Bright and clear

Section 7.2, the Tyndall Beam examination, is very useful for identifying visible contaminants present in new oil. A narrow beam of light projected upward through a sample will reflect from particles of metal, carbon, paper, or other material. Cloudiness indicates that water or sludge are present. Valuable information may be obtained by filtering and identifying the residue. If oil is anything other than sparkling clear, examine the sample for sediment and look at the dielectric strength, neutralization number, and IFT test data. Oxidation products can cause cloudiness, but will also cause a high neutralization number and low interfacial tension.

ASTM Test Methods for Pour Point of Petroleum Oils (D 97)

Limit: $-40°C$, max

Many sites within the Northwest United States can experience wintertime ambients below $-40°C$. The extremely cold sites are located in the Montana Rockies, Southern Idaho, and the Grand Tetons. To my knowledge, the influence of extreme cold on today's naphthenic-based insulating oil has not yet caused an equipment failure. Our practice is to run the pour point test for qualification of vendors, but not for routine acceptance purposes. We are not aware of the pour point of an oil being affected by normal aging in a transformer or other electrical equipment unless the oil becomes contaminated with compounds containing n-paraffins or waxy-like substances.

ASTM Test Method for Flash and Fire Points by Cleveland Open Cup (D 92)

Limit: $145°C$, minimum, flash (fire point limit not given in D 3487)

Because of the possibility of a fire related to failure of electrical equipment, it is important to have knowledge of the flash and fire points of oil. We use the method as a qualification test, but not for acceptance purposes. The flash point test provides adequate information on the flammable properties of oil, and the fire point, a higher temperature, is not usually required. Note 1 in D 92 refers to D 93 [ASTM Test Methods for Flash Point by Pensky-Martens Closed Tester]. By virtue of Note 1, we consider D 93 as binding on suppliers as part of D 92. In other words, we expect new oil to be free of flammable gases (such as methane) that may escape detection by the open cup method.

ASTM Test Method for Aniline Point and Mixed Aniline Point of Petroleum Products and Hydrocarbon Solvents (D 611)

Limit: $63–84°C$, max

This is a qualification test. It is not performed at BPA, however. A contract laboratory would do it for us when needed. Typically, reports from the supplier and information on the type of crude are accepted. The aniline point estimates the total aromatic content and indicates the solvency of the oil for materials that are in contact with the oil. It may relate to impulse and gassing characteristics of oil.

ASTM Test Method for Saybolt Viscosity (D 88)

Limit: 3.0 cSt at 100°C, 12.0 cSt at 40°C, 76.0 cSt at 0°C, max

BPA does not perform test D 88 (see D 445 below). Viscosity of oil influences heat transfer rates and consequently temperature rise of apparatus. Viscosity of oil also influences the speed of moving parts in circuit breakers or tap changers, for example. Oil viscosity controls oil treatment rates as well. High viscosity oils are less desirable, especially in cold climates.

ASTM Test Method for Kinematic Viscosity of Transparent and Opaque Liquids (and the Calculation of Dynamic Viscosity) (D 445)

Limit: 3.0 cSt at 100°C, 12.0 cSt at 40°C, 76.0 at 0°C, max

Viscosity measurements are made for qualification purposes. D 445, measuring flow through an orifice, is used as a referee method.

ASTM Test Method for Density, Relative Density (Specific Gravity), or API Gravity of Crude Petroleum and Liquid Petroleum Products by Hydrometer Method (D 1298)

Limit: 0.91 max, at 15°C

Specific gravity measurements are made for acceptance. Specific gravity is important where there is a concern about water in oil freezing and rising to float on top of the oil. A lower limit for specific gravity would eliminate this potential problem. The BPA Laboratory uses a PAAR digital density meter, based on a different measuring principle for routine testing. Normally, density is checked at room temperature. The PAAR meter results compare favorably with D 1298 results.

Design Tests

ASTM Test Method for Coefficient of Thermal Expansion of Electrical Insulating Liquids of Petroleum Origin, and Askarels (D 1903)

This is considered as a design test with little importance to transformer and circuit breaker applications. The test is used for oil-filled cable installations at BPA.

ASTM Test Method for Thermal Conductivity of Liquids (D 2717)

D 2717 test results are useful in engineering calculations relating to the manner in which a given system reacts to thermal stress. The thermal conductivity test measures temperature change across a liquid sample for a known amount of energy input. The test is not practical for acceptance or qualification; it is a design test.

ASTM Test Method for Specific Heat of Liquids and Solids (D 2766)

This standard method is also basically a design test, measuring capacity of oil to absorb thermal energy (amount of energy input to cause a temperature change). The D 2766 test is not essential for qualification or acceptance purposes.

Additional Tests Not in ASTM D 3487

ASTM Method for Analysis of Gases Dissolved in Electrical Insulating Oil by Gas Chromatography (D 3612)

Refining method by-products can include hydrogen and methane. To find out what is being purchased, a dissolved gas analysis test is an appropriate acceptance test. Dissolved gas is removed from oil during normal treatment (drying, dehydrating, and degassing) before being put in transformers. But, the test can be very revealing of unique manufacturing circumstances. On large transformers received without oil, it is of interest to make sure any residual oil left after factory tests is free from combustibles.

Static Charging Tendency

This utility is actively following research and development in this area and is interested in seeing a standard method for evaluating the ability of liquid insulations to generate and transport entrained charge.

ASTM Method for Analysis of Polychlorinated Biphenyls in Insulating Liquids by Gas Chromatography (D 4059)

Limit: 2 ppm, max (lower limit of detection)

Movement to add D 4059 to D 3487 is concurred with. It is important to underscore the legislated requirement restricting commerce of PCB-containing electrical equipment. BPA's current specification requires oil with 2 ppm or less PCB as tested by D 4059 anyway.

Dielectric Strength of Oil in Motion

No standard test method is available for evaluation of the dielectric strength of insulating liquid moving in restricted channels using dissimilar wall materials. Many transformers have to force oil flow with pumps to achieve cooling. Static electrification research reveals that the dielectric strength of oil flowing in regions of electrical stress is a function of flow rate, field strength, temperature, moisture content, and particulate contamination.

ASTM Test Method for Specific Resistance (Resistivity) of Electrical Insulating Liquids (D 1169)

It may be appropriate to add D 1169 to D 3487. The resistivity of oil depends on many factors, including temperature, moisture, and contaminating particles. Some European transformer manufacturers have specified oil requirements to include this or a similar test. For many years, resistivity has been used by cable and capacitor manufacturers as an insulating fluid qualification and acceptance test. Comparing different oils to be used in forced cooled transformers for this characteristic would be beneficial. Test Method D 1169 has special relevance for oil to be used in HVDC equipment.

Bibliography

BPA Invitation for Bids No. DE-FB79–86BP60194, "Insulating Oil," Bonneville Power Administration, Vancouver, WA.
BPA Substation Maintenance Standard No. 6476–000-2, "Insulating Oil," Bonneville Power Administration, Vancouver, WA.
Doble Transformer Oil Purchase Specification (TOPS), revised, Doble Engineering Co., Watertown, MA.

1986 Annual Book of ASTM Standards, Vol. 10.03, ASTM, Philadelphia, PA.
Reference Book on Insulating Liquids & Gases, Section III: *Insulating Oil Test Guide,* Doble Engineering Co., Watertown, MA.

DISCUSSION

C. L. S. Vieira[1] *(written discussion)*—I understand from your presentation that you use ASTM D 3847 (all tests) for the qualification of a new supplier of transformer oil and only repeat some of the tests as a criteria for acceptance. My question is divided into two parts:

1. Are the tests in ASTM D 3487 enough to evaluate a new oil (for example, a European paraffinic oil)?
2. If no, what additional tests do you recommend to evaluate a new product?

D. L. Johnson (author's closure)—Evaluation of a new oil would include all of the D 3487 tests plus any other standard tests considered important by the oil user. For example, as noted in the paper, Standard Test Methods D 1169 and D 3612 could provide results useful in comparing new products to familiar products. The application for the new oil should be considered. ASTM Test Method for Compatibility of Construction Material with Electrical Insulating Oil of Petroleum Origin (D 3455) is suggested to verify that the oil and other material in contact do not change the electrical, physical, or chemical properties of each other.

[1]CEPEL, P.O. Box 2754, Rio de Janeiro, Brazil.

Henry A. Pearce[1]

Significance of Transformer Oil Properties

REFERENCE: Pearce, H. A., **"Significance of Transformer Oil Properties,"** *Electrical Insulating Oils, STP 998,* H. G. Erdman, Ed., American Society for Testing and Materials, Philadelphia, 1988, pp. 47–54.

ABSTRACT: Transformer oil is a refined mineral oil obtained from the fractional distillation of crude petroleum. It is then treated to remove impurities and to obtain the most desirable properties to make it suitable as an insulating and cooling liquid.

The most important property for an insulating liquid is the high dielectric strength and for a cooling liquid the low viscosity. Other properties are important to show the absence of impurities and to show the degree of stability for long life. Properties indicating purity are also important in the consideration of the ability of the oil to be compatible with the other materials used in transformer construction.

The primary enemies of the transformer oil are oxidation, contamination, and excessive temperature. Processing and transformer construction are designed to minimize the effect of these enemies, and tests are conducted to evaluate these considerations.

ASTM Specification for Mineral Insulating Oil Used in Electrical Apparatus (D 3487) has been designed to cover the properties of transformer oil for use as an insulating and cooling medium. Quoting its scope: "This specification is intended to define a mineral insulating oil that is functionally interchangeable and miscible with existing oils, and is compatible with existing apparatus and with appropriate field maintenance, and will satisfactorily maintain its functional characteristics in its application in electrical equipment."

Much effort is expended by manufacturers to obtain and maintain quality oil for use in their equipment.

KEY WORDS: transformer oil, insulating oil, cooling liquid, dielectric liquid

Transformer oil is truly a unique material for doing the job for which it is famous, that is, insulating and cooling transformers. Our friend George Westinghouse first built a transformer 100 years ago using oil, copper, cardboard, and steel. And after 100 years of outstanding materials developments we are today building most of our transformers using oil, copper, cardboard, and steel. So, while there have been many changes in crudes, processing, additives, and extractions, the basic insulating oil has a very long history of reliable performance. Actually we are talking about insulating oil rather than transformer oil because it is used in other electrical apparatus such as circuit breakers, capacitors, bushings, etc., as well as transformers.

From the viewpoint of the electrical equipment manufacturer, insulating oil is a refined mineral oil obtained from the fractional distillation of crude petroleum. It is free from moisture, inorganic acid, alkali, sulfur, asphalt, tar, vegetable oils, or animal oils. It offers the following advantages:

1. High dielectric strength.
2. Low viscosity.
3. Freedom from inorganic acids, alkali, and corrosive sulfur.
4. Good resistance to emulsification.
5. Freedom from sludging under normal operation.
6. Rapid settling of arc products.

[1]Manager, Insulating Materials, Westinghouse Electric Corp., Sharon, PA 16146.

47

7. Low pour point.

8. High flash point.

Insulating oil has few enemies, and the manufacturers and users make a concerted effort to protect it from its enemies.

1. *Oxidation*. Oxidation is the most common cause of oil deterioration. The transformer manufacturer puts forth a significant effort to make sure that the tank or case is well sealed from the atmosphere. Careful drying and vacuum processing is conducted to remove air and moisture prior to sealing of the tank after which it is filled with dry air or nitrogen to minimize the exposure to oxygen.

2. *Contamination*. Moisture is the chief among potential oil contaminants. Its presence can provide a source of reactive products with the oil in the presence of heat. It also tends to lower the dielectric properties of the insulating oil. The same precautions that are exercised to protect the oil from oxidation are used to protect it from moisture.

3. *Excessive temperature*. Excessive heat is an enemy of oil. It will cause decomposition of the oil itself and/or it will increase the rate of oil oxidation. The best way to protect against excessive heat is to avoid overload of the transformer.

4. *Corona discharges*. Sparking and local overheating can also break down the oil molecule producing gases and water, which can lead to the formation of acids and sludge.

The properties of insulating oil are very important to the manufacturer of electrical apparatus. Therefore, much time and effort are spent testing the oil and following its properties at the various stages in the design, manufacture, and life of a transformer or other piece of electrical equipment.

Each manufacturer has their own specification for the desired properties of insulating oil. Historically, there were several properties that all manufacturers agreed on and there were some, such as oxidation stability, that involved the use of widely divergent test procedures. In recent years the manufacturers of the electrical apparatus, the producers of insulating oil, and the users of the apparatus have worked together to create one specification in which everyone could agree. This is ASTM Specification for Mineral Insulating Oil Used in Electrical Apparatus (3487). Almost all manufacturers and users have now patterned their requirements after D 3487.

The functional property requirements of mineral insulating oil as listed in D 3487, as well as some typical limiting values, are reproduced in Table 1.

All of the listed properties are important to the equipment manufacturer. Some have higher levels of importance than others. Also, the properties have significance for different areas of consideration. Some properties are used for design calculations, some are used to indicate the uniformity of the oil from the producers, and some are used to reveal normal or abnormal operation of the apparatus in service.

Significance of Test

Dielectric Breakdown ASTM D 877 AND ASTM D 1816, 60 Cycle

The dielectric breakdown voltage is a very important measurement of the electrical stress which an insulating oil can withstand without failure. This is an important design parameter and is used by the design engineer to calculate permissible clearances. Therefore, the manufacturers must observe that the value for the oil being put into the apparatus is maintained at the desired level.

This property is measured by applying a voltage between two electrodes under prescribed conditions under the liquid. There are two ASTM procedures: One is ASTM Test Method for Dielectric Breakdown Voltage of Insulating Liquids Using Disk Electrodes (D 877), which spec-

TABLE 1—*ASTM D 3487.*

The following table lists the physical properties for insulating oil and the appropriate ASTM test numbers for each test procedure.

Property	Specification Value	Typical Value	ASTM No.
Color	0.5 max	0.5	D 1500
Dielectric breakdown at 60 Hz			
0.100″ gap	30 kV min	35 kV	D 877
0.040″ gap	28 kV min	30 kV	D 1816
0.080″ gap	56 kV min	60 kV	D 1816
Neutralization number	0.03 mg max	0.01 mg	D 974
Free or corrosive sulfur	Noncorrosive	Noncorrosive	D 1275
Flash point	145°C (293°F) min	150°	D 92
Pour point	−40°C (−40°F) max	−55°C	D 97
Viscosity			
max. cST at 100°C	3 s max	3 s	D 88
max. cST at 40°C	12 s max	10 s	D 88
max. cST at 0°C	76 s max	70 s	
Moisture content	35 ppm max	20 ppm	D 1533
Specific gravity at 60°F	0.910 max	0.890	D 1298
Inorganic chlorides or sulfates	None	None	D 878
Interfacial tension	40 dynes min	45 dynes	D 971
Power Factor, 60 Hz			
25°C (77°F)	0.05% max	0.01%	D 924
100°C (212°F)	0.30% max	0.10%	D 924
Oxidation stability			
Sludge after 72 h	0.15%	0.1%	D 2440
Sludge after 164 h	0.30%	0.2%	D 2440
Total acid after 72 h	0.5 mg of KOH	0.2 mg of KOH	D 2440
Total acid after 164 h	0.6 mg of KOH	0.3 mg of KOH	D 2440
Aniline point	63 to 83°C	75°C	D 611
Dielectric breakdown, impulse, needle-negative	145 kV	150 kV	D 3300

ifies a test cup equipped with 1-in.-diameter vertical disk electrodes spaced 0.100 in. apart and is reported as the standard test value. It has the longest history of the two procedures and is used for acceptance of oil from the producer. The second procedure is ASTM Test Method for Dielectric Breakdown Voltage of Insulating Oils of Petroleum Origin Using VDE Electrodes (D 1816), which specifies a test cup equipped with spherical electrodes spaced either 0.040 or 0.080 in. apart. This cup includes a stirrer and thus is sensitive to small amounts of contaminants. It is designed for evaluation of new degassed and dehydrated oil and therefore is a good measuring stick of the processing of the oil and the apparatus in the manufacturing location. Figure 1 is a Baur Dielectric Tester setup for D 877 test. Figure 2 is the same tester setup for D 1816.

Dielectric Breakdown, Impulse, ASTM D 3300

This test, ASTM Test Method for Dielectric Breakdown Voltage of Insulating Oils of Petroleum Origin Under Impulse Conditions (D 3300), covers the determination of dielectric breakdown voltage of insulating oil under impulse conditions. The electrodes consist of a brass or steel sphere and a steel point. Electrode configuration may be specified sphere to sphere and point to sphere. This information is important to the designers of equipment as it indicates the

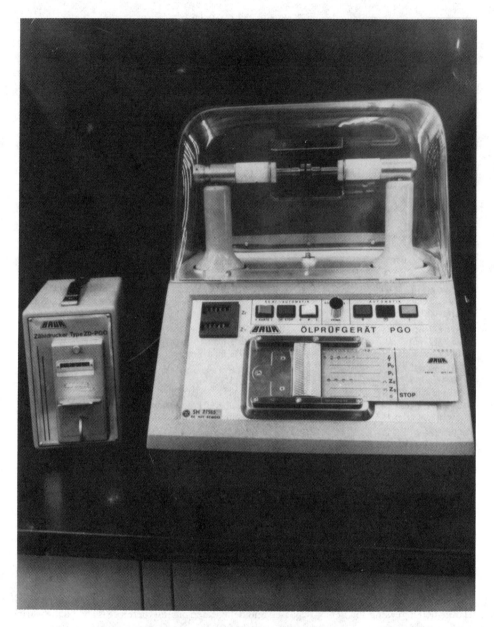

FIG. 1—*Baur dielectric tester setup for D 877 test.*

ability of the oil to withstand transient voltage stresses from such causes as nearby lightning strokes or high-voltage switching operations.

Dissipation Factor (or Power Factor), ASTM D 924

The power factor is the ratio of the power dissipated in the oil in watts to the product of the effective voltage and current in volt-amperes, when tested with a sinusoidal field under pre-

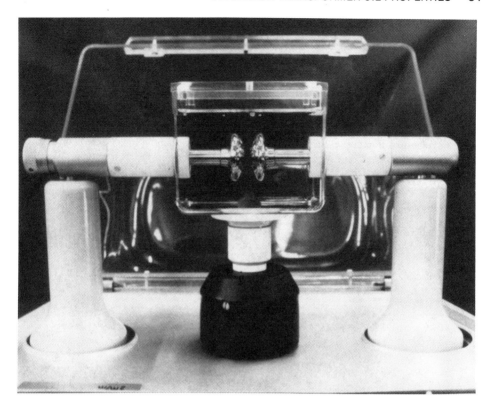

FIG. 2—*Baur dielectric tester setup for D 1816 test.*

scribed conditions. This is a good measuring stick for oil quality. A high power factor value is an indication of the presence of contaminants or deterioration products. (See ASTM Test Method for A-C Loss Characteristics and Relative Permittivity (Dielectric Constant) of Electrical Insulating Liquids (D 924).

Moisture Content, D 1533

The presence of free water may be observed by visual examination in the form of separated droplets or as a cloud dispersed throughout the oil. Water in solution is normally determined by chemical means and is measured in parts per million. Water is important because increased quantity normally decreases the dielectric strength of the oil. The moisture content is important to the manufacturer because it is an indication of the quality of the processing and drying that has been performed on the transformer. It is important in the field operation because increasing values can indicate a breathing tank or the deterioration of cellulose insulation. [See ASTM Test Method for Water in Insulating Liquids (Karl Fischer Method) (D 1533).]

Neutralization Number, ASTM D 974

The neutralization number is the number of milligrams of potassium hydroxide required to neutralize the acid in 1 g of oil. It measures the acid content of the oil. This is most important in indicating chemical change or deterioration of the oil in a unit in service. This normally indicates abnormal operation. Increased acidity may indicate the need to reclaim or replace the oil. [See ASTM Test Method for Neutralization Number by Color Indicator Titration (D 974).]

Interfacial Tension, ASTM D 971

The interfacial tension between the insulating oil and water is a measure of the molecular attractive force between the unlike molecules and is expressed in dynes per centimeter. The test provides a means of detecting soluble contaminants and products of deterioration. It is used by the equipment manufacturer as an indication of the level of contamination of new oil or oil after it has been put into a unit. Figure 3 shows an interfacial tension test being done. [See Test Method for Interfacial Tension of Oil Against Water by the Ring Method (D 971).]

Color, ASTM D 1500

The primary significance of color is to observe darkening in short periods of time, which may indicate either contamination or deterioration of the oil. The color is determined by means of transmitted light and is expressed by a numerical value based on comparison with a series of color standards. [See ASTM Test Method for ASTM Color of Petroleum Products (ASTM Color Scale).]

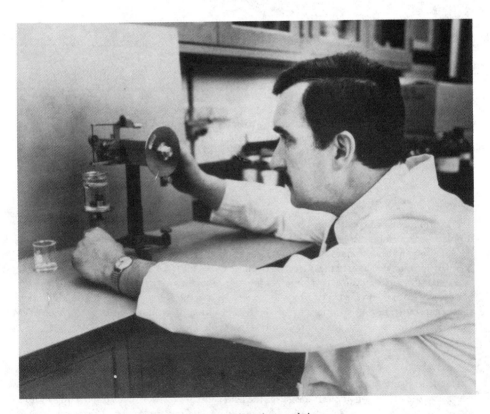

FIG. 3—*Interfacial tension test being run.*

Viscosity, ASTM D 88

Viscosity of insulating oil is one of the significant parameters in determining the heat transfer capability of the fluid. This information, of course, is very important to the equipment designer. The viscosity is the resistance of oil and its resistance to uniformly continuous flow without turbulence inertia and other forces. It is usually measured by timing the flow of a given quantity of oil under controlled conditions. A viscosity increase in an operating unit could indicate deterioration of the oil and formation of sludge. [See ASTM Method for Saybolt Viscosity (D 88).]

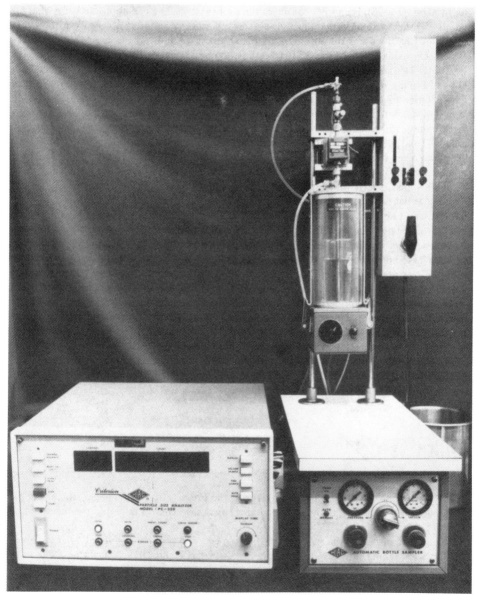

FIG. 4—*HIAC particle analyzer.*

Other Specification Requirements

Other specification requirements are important to characterize a particular oil, but because they are not likely to vary significantly they are not run on a frequent basis by the manufacturer. These include: pour point, ASTM Test Methods for Pour Point of Petroleum Oils (D 97), which is the temperature at which insulating oil will just flow under prescribed conditions. This is important in cold climates to determine start-up conditions. The flash and fire points, ASTM Test Method for Flash and Fire Points by Cleveland Open Cup (D 92), determines the potential fire hazard involved. Specific gravity, ASTM Test Method for Density [Relative Density (Specific Gravity), or API Gravity of Crude Petroleum and Liquid Petroleum Products by Hydrometer Method (D 1298)] is another consideration for heat transfer calculations and also what materials will sink or float in the liquid. Inorganic chlorides or sulfates, ASTM Test Method for Inorganic Chlorides and Sulfates in Insulating Oils (D 878), indicate the degree of treatment of the oil. The oxidation stability, ASTM Test Method for Oxidation Stability of Mineral Insulating Oil (D 2440) or ASTM Test Method for Oxidation Stability of Inhibited Mineral Insulating Oil by Rotating Bomb (D 2112), is an indication of the ability of the natural inhibitors or the added inhibitors to resist the oxidation of the oil.

ASTM standard D 3487 has been designed to cover the properties of transformer oil for use as an insulating and cooling medium. It is presently well structured for this purpose. Quoting its scope: "This specification is intended to define a mineral insulating oil that is functionally interchangeable and miscible with existing oils, and is compatible with existing apparatus and with appropriate field maintenance, and will satisfactorily maintain its functional characteristics in its application in electrical equipment."

A somewhat recent observation that is being used to evaluate oils in units in service is the particle count and the related atomic absorption. The particle count provides the information or the size and number of particles in the oil, and the atomic absorption is a metal-in-oil analysis that tells us what metals are present and their concentrations. Conductive particles are harmful to the dielectric strength of the oil. The generation of metallic particles is usually related to wear and fault problems. Pump bearing wear or arcing and other damage can add metallic particles. This evaluation is increasing in importance in determining what is taking place inside of a transformer. Figure 4 shows a HIAC particle analyzer.

We have not mentioned the determination of PCB content of insulating oil. While this has nothing to do with the functional operations, it is very important to all of us in determining how we may handle, store, transport, or dispose of the oil.

The quality of insulating oil is most important in the operation of electrical apparatus. Therefore, much time and effort is put forth by the producer, the manufacturer, and the user to assure that the desired properties are maintained.

DISCUSSION

C. L. S. Vieira[1] (written discussion)—During the presentation of the paper, the test of gassing tendency was not mentioned as one important test to control the oil in the manufacturer point of view.

I would like discussed the real need of this property and the changes, if any, that the manufacturers are taking into account to design the new EHV and UHV transformers due to the introduction of this new characteristic.

[1]CEPEL, P.O. Box 2754, Rio de Janeiro, Brasil.

Normally, today, the users are looking for gas absorbent oils (negative in ASTM D 2300) but as we saw in the paper by D. L. Johnson, the Bonneville Power Administration accepts oil with values of $+15$ μL/min (Proc. A) or $+30$ μL/min (Proc. B) for all transformers. What is really the actual practice of the manufacturers and/or the utilities concerning this property?

H. Pearce (author's closure)—We as a manufacturer do not test the gassing tendency [ASTM Test Method for Gassing of Insulating Oils Under Electrical Stress and Ionization (Modified Pirelli Method) (D 2300)] on a regular basis. There are many differences of opinion as to what is the most desirable range of values. We have tested each of our suppliers' oils but we do not have this requirement as part of our purchasing specification.

Section III—Analysis of Oil

Jean-Pierre Crine[1]

Newly Developed Analytical Techniques for Characterization of Insulating Oils

REFERENCE: Crine, J.-P., **"Newly Developed Analytical Techniques for Characterization of Insulating Oils,"** *Electrical Insulating Oils, STP 998,* H. G. Erdman, Ed., American Society for Testing and Materials, Philadelphia, 1988, pp. 59–80.

ABSTRACT: A review of newly developed or improved analytical techniques for characterization of insulating oils is made. The emphasis is put on recent techniques, procedures, and applications developed at IREQ in five major areas of oil characterization: thermal aging, electrical properties, particles and metals in oil, gases in oil, and low-temperature applications. It is shown that among other things, the antioxidant content, as measured by HPLC, can be a simple and reliable way to estimate oil oxidation. The influence of particles and metals in oil on a-c breakdown strength are also stressed, and an ultrasonic agitation procedure is proposed to rehomogenize oil samples before electric breakdown measurements. Finally, new trends and future developments, including static electrification and scale model studies, are pointed out.

KEY WORDS: oil characterization, analytical techniques, thermal aging, inhibitor, electric strength, particles, metal in oil, homogenization, gases in oil, paraffins

Electrical engineers dealing regularly with dielectric liquids (mainly oils) need various analytical techniques and methods to evaluate the overall physical, chemical, and electrical properties of these fluids. Over the years a wide variety of tests [see ASTM Specification for Mineral Insulating Oil Used in Electrical Apparatus (D 3487)] have been devised, especially to determine the basic functional property requirements of the liquid (see Table 1). With the development of new dielectric liquids (silicone oil, synthetic fluids, etc.), several of the tests listed in ASTM D 3487 have become obsolete or inadequate. The properties and operation conditions of these new fluids are often very different from those of the conventional mineral oils. This has lead to the development of new or improved analytical techniques for the characterization of insulating liquids.

Dielectric oils are prone to thermal degradation and gas generation. The thermal resistance of any material is generally determined from aging tests relying on accelerated aging conditions [as in ASTM Test Method for Oxidation Stability of Mineral Insulating Oil (D 2440)]. Although these accelerated aging conditions are not representative of the real aging conditions in service, they nevertheless allow comparisons to be made between different materials. There is a need for better (that is more realistic) aging tests, but before these tests be implemented it is necessary to select the analytical techniques that best characterize aging.

Dielectric oils being mostly subproducts of the petroleum industry, their characterization is made more from the chemical than from the electrical point of view (see for example the properties listed in Table 1). With the use of new liquids and the reduction in physical size of modern high-voltage apparatus, the electrical industry has become aware of the necessity to better define its selection criteria for insulating liquids. Their specific behavior under high voltage has also received more attention, and specifications are now more stringent for parameters that do not appear in Table 1, such as gassing, partial discharges, particle concentration, etc. One

[1]Senior research scientist, Institut de recherche d'Hydro-Quebec, Varennes, Quebec, Canada JOL 2PO.

TABLE 1—*Functional property requirements for oils
(ASTM D 3487).*

Property	ASTM Test Method
CHEMICAL	
Oxidation stability	D 2440 or D 2112
Oxidation inhibitor content	D 1473 or D 2668
Corrosive sulfur	D 1275
Water content	D 1315 or D 1533
Neutralization number (acidity)	D 974
PHYSICAL	
Aniline point	D 611
Color	D 1500
Interfacial tension	D 971
Pour point	D 92
Specific gravity	D 1298
Viscosity	D 445 or D 88
Visual examination	D 1524
ELECTRICAL	
Dielectric breakdown, at 60 Hz	D 877 or D 1816
Dielectric breakdown, impulse	D 3300
Power factor	D 924

objective of this paper is to present new or improved analytical techniques for a better characterization of insulating oils. This presentation is arranged along five major themes:

1. Thermal aging characterization.
2. Electrical properties characterization.
3. Particles and metals in oils.
4. Gases in oils.
5. Characterization techniques for low-temperature applications.

The third item is just beginning to receive the attention it deserves by the industry and engineers, and the last one addresses to a specific application.

Another objective of this paper is to review some of the analytical techniques and methods developed at IREQ during the last 15 years. As the research arm of a utility (Hydro-Quebec), a lot of efforts have been devoted in our laboratory to the development of more reliable methods for the evaluation of transformer oil aging. Thus, the emphasis here is put on mineral oil characterization, although the techniques discussed could also be used with other insulating liquids.

Thermal Aging Characterization

New oils can be characterized by their interfacial tension, acidity number, density, and various other parameters (see Table 1). However, as informative as these values can be, they cannot predict the oil lifetime in service, especially for reclaimed oils. A certain amount of new information, such as the molecular weight distribution, nature of aromatics, etc., can be gained by high pressure liquid chromatography (HPLC) analysis [1]. This technique can also be used to evaluate some specific transformations (polymerization, cracking, and unsaturation) during severe electrical and thermal aging [1]. One of the most important application of HPLC with dielectric liquids is to estimate the antioxidant content of inhibited oils. It is known that adding di-t-butyl para cresol (DBPC) to new oils will prolong considerably their oxidation induction time [2]. The estimation of DBPC content in new oil can be done by infrared (IR) spectroscopy

[ASTM Test Method for 2,6-Ditertiary-Butyl Para-Cresol and 2,6-Ditertiary-Butyl Phenol in Electrical Insulating Oil by Infrared Absorption (D 2668)]. However, the technique is not very sensitive, and oxidation by-products absorb IR light at the same wavelengths, which means that it is extremely difficult to evaluate the antioxidant content as a function of aging time. A HPLC procedure developed at IREQ allows the detection of DBPC to content as low as 50 ppm (see Fig. 1) [3]. In fact, this method has also been used with success to determine the antioxidant content in lubricating oils. The possibility of developing a commercial instrument based on this technique is currently examined.

Another technique developed at IREQ to evaluate DBPC content in mineral oil is differential impulsional cyclovoltametry [4]. This technique is less precise than HPLC for DBPC content lower than ≈ 200 ppm and is also more time consuming.

Knowledge of the antioxidant content (if any) in mineral oil is very important because it was shown that its depletion corresponds to a drop of interfacial tension value, a rapid increase of acidity number, and sludge formation [5,6]. This critical point, called the oxidation induction time, depends very much on the chemical structure of the oil and on the antioxidant amount. The fact that the oxidation induction time varies very much with the oil nature implies that considering only the results of measurements performed after 72 and 164 h of accelerated aging at 110°C, as suggested by ASTM D 2440, could possibly be misleading. The 72 and 164 h-values were originally selected because it was thought that an inhibited mineral oil will not be oxidized before 72 h and will be completely oxidized after 164 h. However, some inhibited oils (especially the new ones, as shown in Fig. 2) are oxidized before 72 h, which makes the compari-

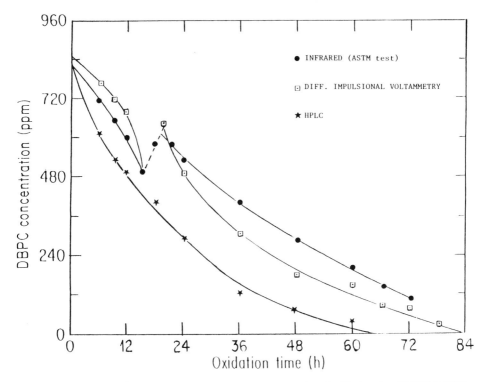

FIG. 1—*Variations of DBPC content in paraffinic oil as a function of oxidation time (ASTM D 2440) as measured by three different techniques; note that IR spectroscopy does not allow the detection of contents smaller than = 120 ppm.*

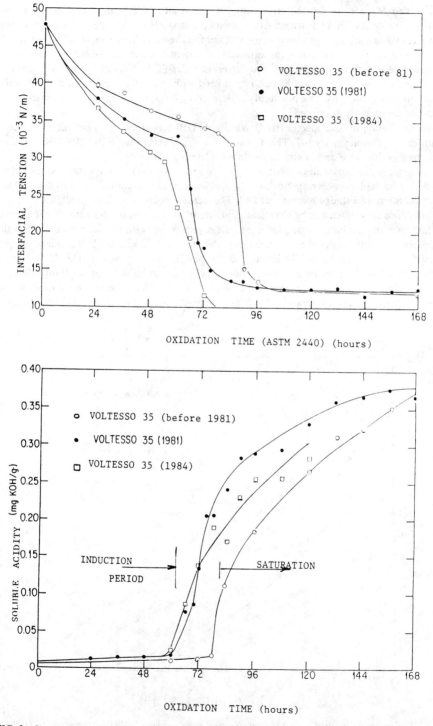

FIG. 2—*Interfacial tension and soluble acidity of samples of a naphthenic oil manufactured on different years as a function of oxidation time (ASTM D 2440). The abrupt variations correspond to sludge formation.*

son between the results at 72 and 164 h worthless. It is therefore suggested to oxidize several oil samples at various aging times, as shown in Figs. 1 and 2, rather than oxidizing two samples for 72 and 164 h. Thus, one can distinguish three phases in the oxidation process (Fig. 2): (1) an induction period which ends when there is no more antioxidant left in the oil; (2) the beginning of oxidation where acidity increases and sludge and peroxides are formed; and finally (3) a saturation period where acidity has reached a maximum value. Note that it is not necessary to measure all these parameters (acidity, peroxides, sludge, antioxidant) to determine the onset of oxidation. The measurement of antioxidant content by HPLC is sufficient to give precise and reliable estimates.

Another method to determine the oxidation onset is to measure the acidity number [ASTM Test Method for Neutralization Number by Color Indicator Titration (D 974)], and the beginning of oxidation corresponds to a rapid increase of soluble acidity (Fig. 2). Note that in aged oil it is difficult to evaluate the neutralization point by visual observation because of sludge formation, particles, and oil darkening. We have improved the sensitivity of the ASTM test method by determining photometrically the neutralization point with a Sybron-Brinkman Digital Probe Colorimeter Model PC800 equipped with a 660-nm filter.

The onset of oxidation corresponds also to the formation of polar products which are the precursors of acid groups. These polar products get unnoticed by standard techniques such as IR spectroscopy, interfacial tension, and acidity number. However, these products are extremely detrimental to the chemical and thermal resistance of the oil (more details in Refs 5–8). The polar products can be separated from the aromatic constituents of the oil and their content estimated in relative units by reverse-phase HPLC (Fig. 3) [7]. Elution of oxidized oil on Fuller's earth removes acidity products, sludge, and particles. However, it does not remove these polar products, and their presence in reclaimed oils (in which 3000 ppm of DBPC was added) explain the poorer thermal resistance of these oils [6–8]. In fact, it was possible to demonstrate that the best time for oil reclaimation corresponds to antioxidant depletion [6–8]. This conclusion would have not been made possible without the use of HPLC for the determination of antioxidant and polar products content.

Finally, an alternative method of evaluating oil thermal stability is to measure the oxidation induction time (OIT) by differential scanning calorimetry (DSC). The OIT is the time where the heat of reaction at a given high temperature ($>150°C$) increases rapidly. The technique is routinely used for the characterization of lubricating oils [9], and it could also be used to characterize insulating oils. The measurement should be made under a high-pressure (500 psi or more) of O_2 in order to reduce the OIT.

Electrical Characterization of Oils

It is well known that it is difficult to characterize dielectric liquids (and especially aged ones) from only their electrical properties. Nevertheless, electrical measurements such as dielectric losses, d-c conductivity, inception voltage for partial discharges, and dielectric strength are often the ultimate criterion for selection of new oils by electrical engineers. We have recently shown that the tan δ and d-c conductivity of mineral oils aged under accelerated conditions depend essentially on three parameters: (1) the dissolved copper content; (2) the peroxides content; (3) the soluble acidity content [5]. Empirical equations fitting the experimental results obtained at 22°C with naphthenic oil (see Figs. 4 and 5) were [5]:

$$\sigma = (1.5 \times 10^{-14} \times C_{Cu}) + (10^{-12} \times C_{SA}) + (1.5 \times 10^{-12} \, C_p) \text{ in } (\Omega \text{ cm})^{-1}$$

$$\tan \delta (60 \text{ Hz}) = (10^{-3} \times C_{Cu}) + (2 \times 10^{-2} \, C_{SA}) + (10^{-2} \, C_p)$$

where C_{Cu}, C_{SA} and C_p are the copper, soluble acidity, and peroxides content of the oil, respectively. These equations suggest that some electrical properties could be predicted from the deg-

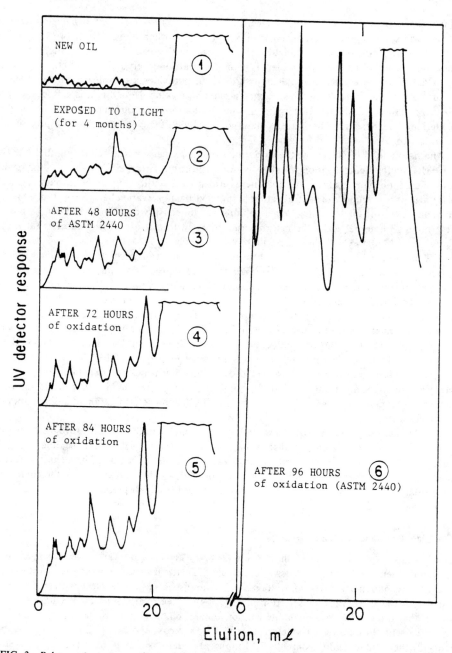

NEW OIL ①

EXPOSED TO LIGHT
(for 4 months) ②

AFTER 48 HOURS
of ASTM 2440 ③

AFTER 72 HOURS
of oxidation ④

AFTER 84 HOURS
of oxidation ⑤

AFTER 96 HOURS ⑥
of oxidation (ASTM 2440)

UV detector response

Elution, mℓ

FIG. 3—*Polar products detected by reverse-phase HPLC in transformer oil aged under various conditions. The UV-response scale is the same for all samples.*

radation products content or vice versa. However, in-service mineral oils are used under a wide variety of conditions and in the presence of impurities, paper, particles, etc. The above equations are therefore less useful, although the electrical properties of oils aged in service depend very much on acidity and on dissolved copper (as shown later).

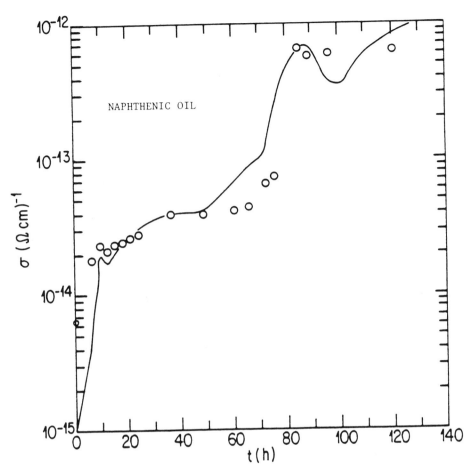

FIG. 4—*Variation of d-c conductivity (22°C) as a function of oxidation time (ASTM D 2440). Experimental results: solid line; calculation (see text): open circles.*

Another dielectric test commonly used in the electrical industry is the so-called "PFVO Doble test," where the dielectric losses of an oil sample heated in air at 95°C are recorded as a function of time [10]. Again, it is extremely difficult to correlate the results thus obtained with real in-service conditions. However, it can be used to compare different new or aged oils.

A major difficulty in the comparison of tan δ or conductivity values of different oils (or different test methods) is the influence of the applied electric field. At low and moderate a-c field intensities, tan δ is nearly constant, but it increases rapidly at high field intensity. Although ASTM D 924 specifies that measurements should be made at voltage stress larger than 2 kV/cm, most results are obtained with a few applied volts per cm and only few are obtained with fields of several kV/cm. Aged oils with their conductive impurities are much more sensitive to field intensity than new oils. In our opinion, the measurements of electrical properties should always be made at, at least, three field intensities ranging from a few volts per cm up to field values similar to those encountered in service. That would certainly reduce the large scatter of values often reported by various laboratories (and therefore various experimental conditions).

Equipment and test methods for electrical properties measurements have not been greatly modified over the last 20 years. One promising and recent development is a complex permittiv-

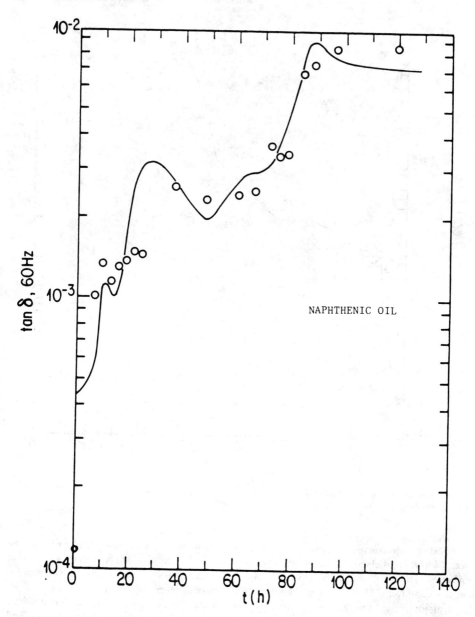

FIG. 5—*Variation of tan δ (60 Hz, 22°C) as a function of oxidation time (ASTM D 2440). Same symbols as in Fig. 4.*

ity measurements system called Microdielectrometry and manufactured by Micromet Instruments, Inc. [11]. The original feature of the instrument is that it uses coplanar interdigital (comb-like) electrodes deposited on a silicon chip which is acting as a sensor. The small size of the sensor and the fact that electrodes are coplanar mean that small liquid samples are required. The sensor could also be used in locations that would have been otherwise unaccessible for conventional parallel disk electrodes. Finally, since the floating electrode on the sensor is a charge integrating device, there is no lower limit of frequency. This system seems well suited for very low-frequency measurements (down to $\approx 10^{-3} - 10^{-4}$ Hz) where ionic currents and interfacial phenomena can then be estimated with more accuracy. Note that such low-frequency measurements can now be relatively easily achieved with the recent development and use of frequency response analyzers [12].

The detection of partial discharges and the significance of their inception voltage are still a matter of debate [13,14]. In fact, the major problem is that there is yet no standard experimental procedure and no agreement on what should be detected: Are the apparition of bubbles, light [15], or sound emission [16] the characteristics of partial discharges? Another experimental difficulty is the nature of electrodes: Metallic or metal covered with paper or any insulating material? [14] Much remains to be done in the basic understanding of the phenomena underlying partial discharges and much more before a rational and reliable interpretation of the results can be done. However, great progress has been made in their detection, mainly due to the extensive use of computers [17].

Electric breakdown strength is one of the most important parameter characterizing an insulating liquid. However, it is very difficult to correlate its value with a given lifetime in service. This is essentially due to the fact that breakdown is associated with many parameters (water, gases in oil, sampling conditions, particles, etc.). The following section is devoted to particle characterization and their influence on electric breakdown measurements.

Particles and Metals in Oils

It is well known that solid particles in suspension in insulating liquids affect their dielectric strength depending on their concentration, size, and nature [18-20]. However, up to recently only limited attention has been paid to this problem. This was essentially due to the lack of a simple and fast analytical technique. There is presently no specific ASTM test method devoted to particle counting in insulating liquids, although ASTM Method for Microscopial Sizing and Counting Particles from Aerospace Fluids on Membrane Filters (F 312) could be used. Particle counting with the advent of automatic particle counters has now become a simple, fast, and inexpensive technique. However, it is plagued with experimental difficulties that can make its results absolutely worthless. The major problem is particle deposition at the bottom at the sampling vessel. It was shown that particle deposition in transformer oil is relatively rapid, and in only 1 h all particles larger than 50 μm are deposited [20]. Smaller ones take more time, but, as shown in Fig. 6, particle deposition is very important in only one day, which results in a significant improvement of the a-c breakdown strength of the oil. Note that in agreement with ASTM Test Method for Dielectric Breakdown Voltage of Insulating Oils of Petroleum Origin Using VDE Electrodes (D 1816), the oil container was inverted gently prior to the breakdown test; it was shown elsewhere that this procedure does not restore the original particle distribution [22]. This means that oil samples stored several days before measurements (as is often the case) are not representative of the oil situation at the sampling time. This also means that the oil shown in Fig. 6 had a breakdown voltage (V_b) below the "acceptable" 50-kV value (for 2 mm) just after sampling. However, one day later its V_b was over the critical 50-kV value and it had an even more artificially high V_b value after seven days at rest. In other words, the breakdown voltage value, which is one of the most important routine measurements performed by most electrical utilities, may give a completely misleading indication of the real condition of the oil in the transformer if it is performed several days after sampling. It should be noted that the results shown in Fig. 6 correspond to a particularly dirty transformer oil. Most oil samples taken in transformers in service have particle concentrations of the order of some thousands per 10 c^3.

It is therefore absolutely necessary to rehomogenize oil samples before measurements such as particle count, metal content, dielectric strength, etc. At IREQ, we have selected ultrasonic agitation because it is easy to use and requires low cost equipment. Bubble formation has been occasionally observed during ultrasonic agitation but they disappear very rapidly, and it was

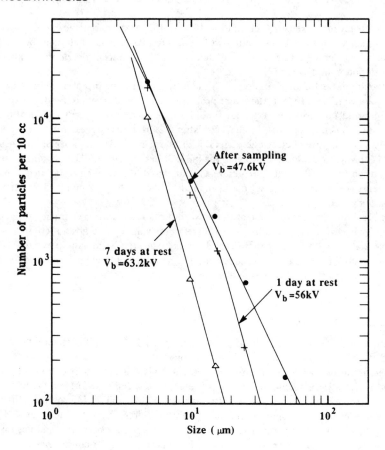

FIG. 6—*Particle distributions in aged transformer oil as a function of resting time. The breakdown voltage (ASTM D 1816) V_b is also indicated.*

verified experimentally that this does not affect particle count and dielectric strength measurements. The oil sample can absorb some water during ultrasonic agitation, but for a relatively dry sample (that is, <15 ppm of water) the water content increase is less than 2 to 3 ppm (as long as the oil is not directly exposed to air). Interestingly, an International Electrotechnical Commission (IEC) working group is currently developing a procedure of oil homogenization also based on ultrasonic agitation [21].

Detailed experimental conditions are given in Ref 22; to summarize, the recommended ultrasonic homogenization is as follows:

1. Fill a small ultrasonic bath (150 W) with 1 L of water.

2. Agitate one oil bottle (1 L or less) for 25 min (do not forget to unscrew the bottle's cap to avoid overpressure induced by liquid expansion).

3. Perform the measurement (particle counting, electric breakdown test, metals content, etc.) immediately afterward.

Particle deposition is not only important for particle counts and dielectric strength measurements but also affects other measurements such as the metal content of the oil. It is well known that metals, especially copper and iron, are dissolved during oil aging [2,5,20]. Figure 7 shows the copper content of some transformer oils measured after various deposition times; it is obvi-

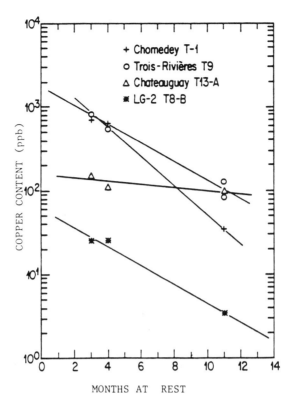

FIG. 7—*Deposition as a function of resting time of copper in samples of oil taken in various transformers in service.*

ous that significant differences are observed and again it could wrongly be concluded that, in some cases, the oil is still acceptable when, in fact, it is aging very rapidly. Thus, homogenization is requested before metals in oil analysis.

In addition to the dissolved copper induced by aging, transformer oil may also contain some copper (and other metal) particles in suspension related to the windings or contact problems [23]. Since most suspended particles can be removed by filtering, important information regarding transformer aging can be gained by separating the two kind of metallic contents. Table 2 shows some typical results obtained with two apparatus where the copper and lead contents are sensitive to filtering whereas the iron and zinc contents are not; the last two metals are therefore essentially dissolved in oil as submicron particles.

The very low metal contents reported in Table 2 affect significantly several electrical properties of oils, as discussed in the section "Electrical Characterization of Oils." They are also extremely difficult to estimate, which explains, in fact, why there are only very few references on that aspect [20,23]. To the best of our knowledge the best analytical technique to measure such low metal contents is atomic absorption spectroscopy (AAS) performed with a graphite furnace. Without a graphite furnace, ordinary flamme AAS or Ion-Coupled Plasma AAS are not sensitive enough (minimum of = 1 ppm for most elements) to detect 10 to 100 ppb of impurities in oil. Another analytical technique that has proven to be reliable for copper content measurements is neutron activation analysis (NAA) [5,20]. This technique can reliably detect very low contents of copper (less than 5 ppb), but it is not very sensitive to iron (more than 1000 ppm are required) and cannot detect lead. By any means, detection of copper by NAA (see Fig. 8) is

TABLE 2—*Influence of filtering on some metal-in-oil contents.*

Oil sample	Copper, ppb	Iron, ppb	Lead, ppb	Zinc, ppb
No. 1				
Before filtering	110	3.7	3.7	<2
After filtering	81	3.7	1.9	<2
No. 2				
Before filtering	105	2.3	8.8	<2
After filtering	69	2.0	3.0	<2

much simpler and more precise than the photometric analysis recommended by ASTM Test Method for Copper in Electrical Insulating Oils by Photometric Analysis (D 2608). Whatever the technique used, sample preparation [see ASTM Test Method for Copper in Electrical Insulating Oil by Atomic Absorption Spectrophotometry (D 3635)] and container cleanliness are crucial steps in obtaining reliable results.

As useful as they are, particle counting and metals in oil analysis cannot give a description of the shape and nature of all particles. In transformers, for example, most particles are long cellulose fibers (whose real length is rarely measured correctly by particle counters) and small (<2 μm) carbon particles. They can be identified by microscopic observation after deposition on a membrane, as suggested in ASTM F 312. However, this is a long and tedious procedure which may be of limited interest for routine measurements. In addition, some particles get through the membrane, and it is not an easy job to distinguish a cellulose fiber on an acetate membrane [24]. These difficulties may be overcome by using ferrography, a technique originally developed to study wear and abrasion particles, where all particles are deposited and permanently fixed on a glass plate under the action of a magnet (see Fig. 9).

Magnetic particles are aligned by the magnetic field and can be easily detected (see Fig. 10) under microscopic observation. The approximate nature of other particles can also be determined by an appropriate choice of various light filters (Fig. 11) [24]. A rapid microscopic observation of the glass plate gives interesting information on the shape and nature of the particles. In addition, particles being permanently fixed on the glass plate can be precisely identified at any time by a technique such as electron microprobe. Thus, ferrography combined to particle counting can give a qualitative estimation of the shape, kind, and nature of particles in an insulating liquid. Note that ferrography glass plates could eventually be analyzed by an image analysis system for automated particle size distribution measurements.

Gases in Oils

When a dielectric liquid is subjected to electric arcs, localized hot spots, or partial discharges, it generates various gases whose chemical composition depends on the nature of the fault. The analysis of dissolved gases has become a very useful method of diagnosis of incipient faults in transformers [25,26]. The IEC is currently reviewing its standards of gas analysis in oil, especially with regards to contents below the ppm range.

A method has been developed at IREQ which addresses this problem. It consists in the preparation of standard oil samples containing volumetrically determined amounts of dissolved gases [27]. This method has been successfully used by Hydro-Quebec to control the quality of the analyses performed by various laboratories bidding for services. It has also been used by IEC to evaluate the repeatability and reproductibility of various gas extraction and chromatographic techniques used worldwide. This has allowed us to show that ASTM Method for Analysis of Gases Dissolved in Electrical Insulating Oil by Gas Chromatography (D 3612) is a very good

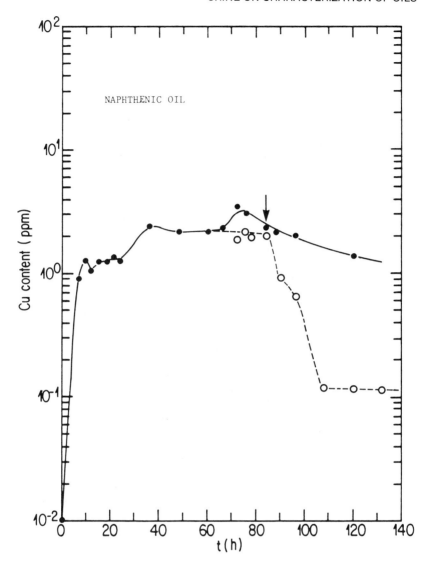

FIG. 8—*Dissolved copper content (as measured by neutron activation analysis) in transformer oil as a function of oxidation time (ASTM D 2440). Filled symbols: unfiltered; open symbols: filtered oil. The arrow corresponds to sludge formation.*

method as compared to those used in Europe and Japan. Note that this applies only to the original method of ASTM D 3612 and not to the modified versions. Special diagnosis methods are used at Hydro-Quebec for the interpretation of gas-in-oil results [26,28], which have been, in several instances, more reliable than standard codes [29].

One of the most dangerous gases generated by partial discharges, arcs, or hot spots is hydrogen. In order to be able to follow continuously the time-evolution of hydrogen content in HV transformers, a detector was developed at IREQ [30]. Basically, it is an inverted fuel cell, and the presence of hydrogen induces a current proportional to the gas content. Commercial instruments called Hydran 101 and 201 (Fig. 12) have subsequently been marketed by Syprotec Inc. [31]. To the best of our knowledge this hydrogen detector has proven satisfactory to users [32].

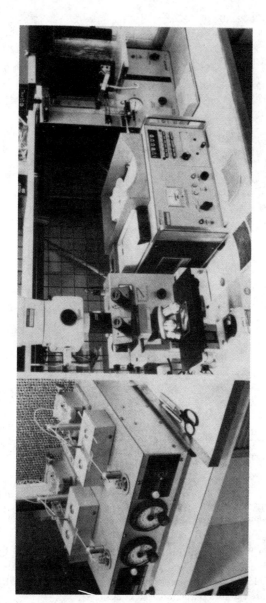

FIG. 9—*Photos of the ferrograph, the microscope, and the automatic article counter used at IREQ.*

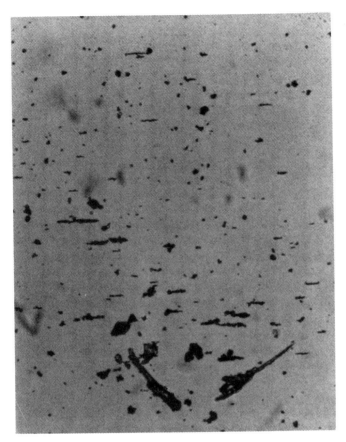

FIG. 10—*Photograph of particles in aged oil deposited on a ferrogram. Note the magnetic particles aligned parallel to the magnetic field. Other metallic particles and fibers are also visible. Magnification: ×100.*

It should be noted that the hydrogen analysis should be complemented by a more complete gas analysis to determine the origin of the fault.

For transformer applications, the tendency is to develop a simple and inexpensive gas monitor that could be installed in or near the apparatus. Westinghouse Electric Corp. is currently developing an in-situ gases detector to monitor power transformers [33]. A multigas in-situ monitor is also under study at IREQ [34], and one Japanese team has recently presented results obtained with a new and inexpensive gases detector [35].

Characterization Techniques for Low Temperature Applications

The dwindling reserves of mineral naphthenic oil have stirred a continuous interest for other types of mineral liquids, especially for the so-called paraffinic oils. Paraffinic and naphthenic oils have essentially the same electrical and chemical properties [36]. However, oils with larger paraffinic content have significantly larger viscosities at low temperatures. It was shown that a paraffinic oil made from Canadian crude was completely solid at −23°C [37]. The wax formation was also observed in oils of various origins, and it was determined that it occurs every time that the mineral oil contains more than 4% of linear paraffins [38]. This solidification problem

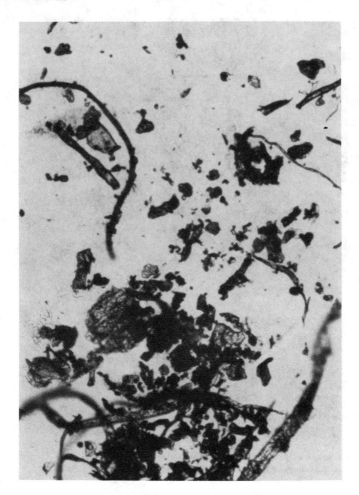

FIG. 11—*Photograph of the ferrogram of another transformer oil sample taken in a transformer in service. Long fibers and carbon particles are the major contaminants. Note the wide variety of particle size and shape. Magnification: ×100.*

of paraffinic transformer oil could be disastrous in a cold weather country. It is therefore important to evaluate the paraffins content in mineral oils. This can simply be done by measuring the cloud point or the pour point as a function of paraffins content [38]; these are not very precise methods. We obtained much more precisely the paraffins content of mineral oils by differential scanning calorimetry (DSC) (Fig. 13) or by X-ray diffraction measurements [38]. The latter method is less precise for low wax content but it has the advantage of characterizing the nature of paraffins (linear or amorphous).

A complete characterization of low-temperature liquids should, of course, include viscosity measurements. This can be done with commercial viscosimeters such as the Brookfield or by the so-called Stoke method [37].

It is possible to improve, that is, to reduce, the viscosity of oils at low temperatures by adding a flow-improver. The amount of flow-improver in unaged or aged oil can be precisely evaluated by gel permeation chromatography (GPC) [39].

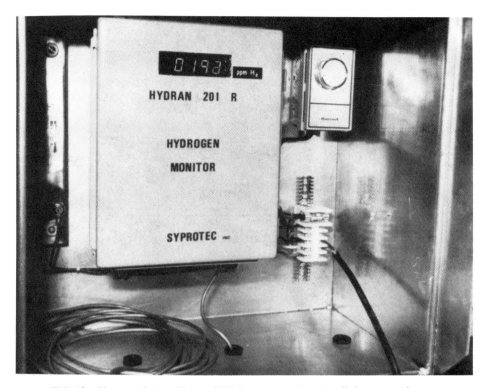

FIG. 12—*Photograph of an Hydran 201R hydrogen monitor installed on a transformer.*

New Trends and Future Developments

Static Electrification

Static electrification in transformer oils has recently gained a renewed interest due to some unexpected and yet unexplained transformer faults [*40*]. When an insulating fluid flows over a conducting component charge, separation occurs and the liquid gets charged. If the voltage thus generated exceeds the breakdown strength of the liquid, discharges will occur. The separation of charges at the solid/liquid interface is affected by several parameters which are not yet all known. In fact, the basic phenomenon of static electrification is poorly understood and is now the subject of several studies all around the world [*40-43*]. There is currently a need for a more detailed standard experimental procedure than the present ASTM test method D 2679 [discontinued—to be replaced by D 4470]. A meaningful definition for the "electrostatic charging tendency" of a liquid is also needed.

Metal Deactivators

It has been known for a long time that adding a metal deactivator to insulating liquids, especially transformer oils, leads to a better thermal resistance [*44,45*]. It was recently shown that it may also reduce the electrostatic charging tendency of oil [*42*]. Research has to be done to determine whether metal deactivators should be really added to transformer and capacitor oils to prolong their lifetime. An analytical procedure to detect its content and variation with time has also to be devised.

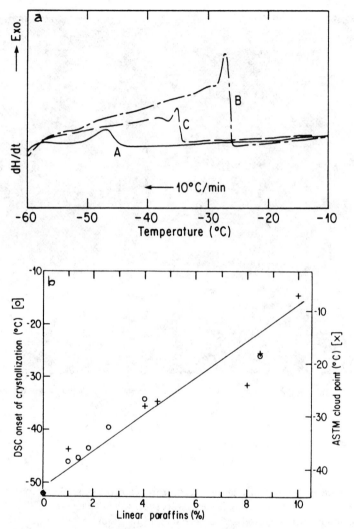

FIG. 13—*DSC evaluation of the paraffins content in transformer oils.* (a) *Typical DSC curves showing the onset of crystallization of naphthenic oil (A), paraffinic oil (B), and partially dewaxed oil (C).* (b) *Influence of the amount of linear paraffins on the onset of crystallization and on the cloud point of various oils.*

Ionic Impurities in Liquids

Although it is agreed that ionic impurities are the major charge carriers in most insulating liquids, there are still few experimental measurements of the ionic contents of these liquids. With the advent of ion chromatographs, one may expect that more data would be soon available. Again, a standard experimental procedure is required since the ionic content could vary greatly with the experimental conditions.

Scale Model Studies

Dielectric liquids are generally used in power equipment such as transformers, capacitors, or cables where they are in contact with various insulating and conducting materials. On the long

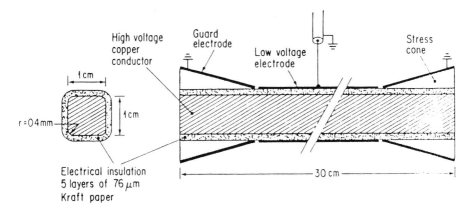

a) Schematic representation of the scale models used by Ref. 46

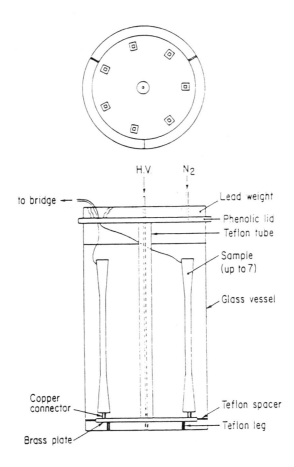

b) Experimental set-up used to test the above scale models.

FIG. 14—*Scale models developed and tested at IREQ to simulate cable and transformer aging.* (a) *Schematic representation of the scale models used by Ref* 46. (b) *Experimental setup.*

TABLE 3—*Some new analytical techniques developed or used at IREQ.*

Technique	Property or Phenomenon Characterized	Complement or Improve ASTM Test
HPLC	Antioxidant content (onset of oxidation)	D 2668, D 1473, D 2446
Reverse-phase HPLC	Polar groups	D 2440, D 974, D 1902
Neutron Activation Analysis	Dissolved copper content	D 3635
Atomic Absorption (with graphite furnace)	Impurity contents	D 3635
Automatic particle counting	Number of particles	F 312, D 1816
Ferrography	Nature and shape of particles	F 312, D 1816
Ultrasonic agitation	Sample homogenization	D 3613, D 1816, D 3635
Differential Scanning Calorimetry	Onset of oxidation (OIT), paraffins content	D 2766
GPC	Flow improver content	
Hydran monitor	Detector of hydrogen	D 3612

term, these materials tend to interact with the liquids either by dissolution, swelling, electrochemical reactions, etc. This means that a complete characterization should take into account the liquid itself and also other parameters or phenomena associated with its environment.

One approach in this direction is to use scale models made with various metals, paper, varnish, etc. (Fig. 14). At IREQ we have developed a scale model that could simulate the materials and conditions in transformers or in oil-filled cables [46]. Scale model aging studies are very useful because they provide data obtained midway between service aging conditions in a full scale apparatus and accelerated laboratory testing, as in ASTM D 2440.

Identification of Oil Constituents by GC/MS

Identification of the various components of mineral oils by gas chromatography/mass spectrometry (GC/MS) is not a new topic. Although some work has started more than 30 years ago, it is still far from being complete, largely because mineral oils are natural and very complex compounds. Development of new experimental procedures and analytical techniques are required before significant progress can be made.

Conclusions

The advantages and limits of a wide variety of newly developed analytical techniques (see Table 3) for insulating liquids have been discussed. New trends and future developments have also been pointed out. One may expect that the current implementation of computer-controlled instruments will eventually lead to significant modifications of laboratory procedures and analyses.

Acknowledgments

It is a pleasure to thank my colleagues C. Lamarre, M. Duval, C. Vincent, D. Couderc, G. Belanger, and R. Malewski for their assistance in the preparation of this review.

References

[1] Duval, M. and Lamarre, C., "The Characterization of Electrical Insulating Oils by HPLC," *IEEE Transactions on Electrical Insulation*, Vol. 12, 1977, p. 340.

[2] Melchiore, J. J. and Mills, I. W., "The Role of Copper During the Oxidation of Transformer Oils," *Journal of the Electrochemical Society*, Vol. 112, 1965, p. 390.

[3] Lamarre, C., Duval, M., and Gauthier, J., "Dosage par chromatographie liquide haute performance du DBPC dans les huiles de transformateur neuves ou usagées," *Journal of Chromatography*, Vol. 213, 1981, p. 481.

[4] Bélanger, G., Lamarre, C., and Crine, J. P., "Antioxidant Analysis in Transformer Oils," Minutes of 48th Annual International Conference Doble Clients, Section 10-601, 1981, Doble, Inc., Watertown, MA.

[5] Lamarre, C., Crine, J. P., and Duval, M., "Influence of Oxidation on the Electrical Properties of Inhibited Naphtenic and Paraffinic Transformer Oils," *IEEE Transactions on Electrical Insulation*, Vol. 22, 1987, p. 57.

[6] Lamarre, C., Crine, J. P., and St-Onge, H., "Antioxidant Functionality in Liquid and Solid Dielectric Materials," CIGRE Symposium on New Materials, Vienna, May 1987.

[7] Duval, M., Lamarre, C., and Giguère, Y., "Reverse-Phase HPLC Analysis of Polar Oxidation Products in Transformer Oils," *Journal of Chromatography*, Vol. 284, 1984, p. 273.

[8] "Study of the Thermal Resistance of Reclaimed Transformer Oils," Report No. 203 T 444, Canadian Electrical Association, Montreal, Canada, 1986.

[9] Walker, J. A. and Tsang, W., "Characterization of Lubricating Oils by Differential Scanning Calorimetry," SAE Technical Paper Series No 801383, Society of Automotive Engineers, Warrendale, PA, 1980.

[10] Doble Engineering Co., 85 Walnut St., Watertown, Mass. 02172 (see also Ref *45*).

[11] Micromet Instruments Inc., 21 Erie St., Cambridge, MA 02139.

[12] See among others J. Pugh, 1984 IEEE Conference Dielectric Materials & Applications, London, No. 239, 1984, p. 247.

[13] "Engineering Dielectrics Vol. 1," in *Engineering Dielectrics, Vol. 1, Corona Measurement and Interpretation, ASTM STP 669*, R. Bartnikas and J. McMahon, Eds., 1979.

[14] Fallou, B., Perret, J., Samat, J., and Vuarchex, P., "Évolution des critères de sélection des liquides isolants," CIGRE Conference, Paper 15-10, 1986.

[15] Pfeiffer, W., "Fast Measurements Techniques for Research in Dielectrics," *IEEE Transactions on Electrical Insulation*, Vol. 21, 1986, p. 763.

[16] Harrold, R. T., "Acoustic Theory Applied to the Physics of Electrical Breakdown in Dielectrics," *IEEE Transactions on Electrical Insulation*, Vol. 21, 1986, p. 781.

[17] James, R. E., Trick, F. E., Phung, B. T., and White, P. A., "Interpretation of Partial Discharge Quantities as Measured at the Terminals of HV Power Transformers," *IEEE Transactions of Electrical Insulation*, Vol. 21, 1986, p. 629.

[18] Oomen, T. V. and Petrie, E. M., "Particle Contamination Levels in Oil. Filled Large Power Transformers," IEEE Power Engineering Society Summer Meeting 1982, IEEE, New York, NY.

[19] Miners, K., "Particles and Moisture Effect on Dielectric Strength of Transformer Oils Using VDE Electrodes," *IEEE Transactions PAS*, Vol. 101, No. 3, 1982.

[20] Vincent, C. and Crine, J. P., "Evaluation of In-Service Transformer Oil Condition from Various Measurements," Conference Record 1986 IEEE International Symposium of Electrical Insulators, IEEE, New York, NY, 1986, p. 310.

[21] IEC SC10A—Special Working Group on Particle Sizing and Counting, 1985.

[22] Vincent, C., Crine, J. P., and Olivier, R. G., "Ultrasonic Homogenization of Particles in Transformer Oil," Doble Client Conference, April 1987, Doble, Inc., Watertown, MA.

[23] Skog, J. E., de Giorgio, J. B., Jakob, F., and Haupert, T. J., "Location of Incipient Faults by Metal in Oil Analysis," Doble Client Conference, 1980, Doble, Inc., Watertown, MA.

[24] Olivier, R. G., Vincent, C., and Crine, J. P., "Analysis of Particles in Transformer Oil by Ferrography," Doble Clients Conference, Sec. 10-401, 1986, Doble, Inc., Watertown, MA.

[25] Baker, A. E., Griffin, P. J., and Locke, C., "An Update on Fault Gas Analysis. A Review," Doble Clients Conference, Section 10-302, 1986, Doble, Inc., Watertown, MA.

[26] Duval, M., "Fault Gases Formed in Oil-Filled Breathing EHV Power Transformers. The Interpretation of Gas Analysis Data," IEEE Power Engineering Society Summer Meeting, Paper C74 476-8, IEEE, New York, NY, 1974.

[27] Duval, M. and Giguère, Y., "Preparation of Standard Samples of Dissolved Gases in Insulating Oils," Doble Clients Conference, Section 10C-01, 1984, Doble, Inc., Watertown, MA.

[28] Dind, J. E. and Régis, J., "How Hydro-Québec Diagnoses Incipient Faults by Using Gas-in-Oil Analysis," *Pulp and Paper Canada*, Vol. 76, 1975, p. 61.

[29] See among others, Rogers, R. R., "IEEE and IEC Codes to Interpret Incipient Faults in Transformers," *IEEE Transactions on Electrical Insulation*, Vol. 13, 1978, p. 349.

[30] Bélanger, G. and Duval, M., "Monitor for Hydrogen Dissolved in Transformer Oil," *IEEE Transactions on Electrical Insulation*, Vol. 12, 1977, p. 334.

[31] Syprotec, Inc., 88 Hymus Road, Pointe-Claire, Qc, Canada H9R 1E4.

[*32*] Sparks, R. D. and Turney, J. H., "Tennessee Valley Authority's Application of the Hydran 201 on-line Hydrogen Monitor to 500 kV Transformers," Doble Clients Conference, Sec. 6-401, 1986, Doble, Inc., Watertown, MA.

[*33*] Nilsson, S. L., "Development of New Techniques for Operation and Maintenance of High Voltage Substations and Substation Equipment," Doble Clients Conference, Sec. 2-301, 1986, Doble, Inc., Watertown, MA.

[*34*] Malewski, R., Douville, J., and Bélanger, G., "Insulation Diagnostic System for HV Power Transformers in Service," 1986 CIGRE Conference, Paper 12-01, 1986.

[*35*] Tsutokia, H., Sugawara, K., Mori, E., and Yamaguchi, H., "New Apparatus for Detecting Transformer Faults," *IEEE Transactions on Electrical Insulation*, Vol. 21, 1986, p. 221.

[*36*] Duval, M., Cauchon, D., Lamothe, S., and Giguère, Y., "Paraffinic Transformer Oils for Use at Low Temperatures," *IEEE Transactions Electrical Insulation*, Vol. 18, 1983, p. 586.

[*37*] Langhame, Y., Castonguay, J., Crine, J. P., Duval, M., and St-Onge, H., "Physical Behavior of Paraffinic Oils at Low Temperatures," *IEEE Transactions on Electrical Insulation*, Vol. 20, 1985, p. 629.

[*38*] Crine, J. P., Duval, M., and St-Onge, H., "Determination of the Nature and the Content of Paraffins in Mineral Transformer Oils," *IEEE Transactions on Electrical Insulation*, Vol. 20, 1985, p. 419.

[*39*] Duval, M., Lamothe, S., Cauchon, D., Lamarre, C., and Giguère, Y., "Determination of Flow-Improver Additives in New and Aged Insulating Oils by Gel Permeation Chromatography," *Journal of Chromatography*, Vol. 244, 1982, p. 169.

[*40*] Griffin, P. J., "Update on Static Electrification in Transformer Oils," Doble Clients Conference, Sec. 6-1201, 1986.

[*41*] Crofts, D., "The Static Electrification Phenomena in Power Transformers," Record of Conference Electrical Insulation Dielectric Phenomena, 1986.

[*42*] Yasuda, M., Goto, K., Okubo, H., Ishii, T., Mori, E., and Masunaga, M., "Suppression of Static Electrification of Insulating Oil for Large Power Transformers," IEEE Power Engineering Society Winter Meeting, Paper 82WM 197-2, 1982, IEEE, New York, NY.

[*43*] Brzostek, E. and Kedzia, J., "Static Electrification in Aged Transformer Oil," *IEEE Transactions on Electrical Insulating*, Vol. 21, 1986, p. 609.

[*44*] Hughes, F. and Haydock, P. T., "The Industrial Use of Passivated Transformer Oil," *Journal of the Institute of Petroleum*, Vol. 50, 1964, p. 239.

[*45*] Duval, M. and Crine, J. P., "Dielectric Behavior and Stabilization of Insulating Oils in EHV Current Transformers," *IEEE Transactions on Electrical Insulation*, Vol. 20, 1985, p. 437.

[*46*] "Life Testing of Catalytically Dewaxed Paraffinic Transformer Oil," Report No. 000 T 453, Canadian Electrical Association, Montreal, Canada, 1985.

DISCUSSION

C. L. S. Vieira[1] *(written discussion)*—The study of thermal aging presented was made with oils that had at least 800 ppm of DBPC. Considering that some oils do not have DBPC added, what will be the evaluation of the decomposition products? In this case the suggested control of remaining DBPC content during the life of the oil when in service will not be useful. What additional control should be added?

J. P. Crine (author's closure)—The most sensitive method to evaluate thermal aging of an insulating oil (with or without antioxidant) is to measure its polar groups content by HPLC (see Refs 7 and *8*). This technique requires relatively sophisticated equipment and a skilled operator. A less expensive, though much less sensitive, method is to measure the interfacial tension; a value lower than 25 to 30 mN/m generally indicates significant thermal degradation of the oil sample.

Finally, note that the so-called noninhibited oils may contain up to 800 ppm of DBPC. Therefore, it may be worth verifying whether some antioxidant is present in the oil before performing other measurements.

[1]CEPEL, P.O. Box 2754, Rio de Janeiro, Brazil.

Section IV—Dissolved Gas in Oil

Leo J. Savio[1]

Transformer Fault Gas Analysis and Interpretation—A User's Perspective

REFERENCE: Savio, L. J., **"Transformer Fault Gas Analysis and Interpretation—A User's Perspective,"** *Electrical Insulating Oils, STP 998,* H. G. Erdman, Ed., American Society for Testing and Materials, Philadelphia, 1988, pp. 83–88.

ABSTRACT: All transformers generate gases during their normal operation as a result of aging and oxidation exclusive of abnormalities. From an operational point of view, it is important to, first, detect and recognize the existance of a problem by observing or detecting deviations from gases considered normal, second, to evaluate the impact on the operation of the transformer (serviceability), and, lastly, to take appropriate action, such as removal from service or increased monitoring frequency. This paper will address these considerations from a utility's perspective using IEEE C57.104 as a guide.

KEY WORDS: transformer, combustible gas, analysis, oil, gases, diagnosis, dissolved gas, IEEE C57.104

Transformers are an important and normally a very reliable component of a power system. Over the years various techniques have evolved to monitor the serviceability of the power transformer, such as testing of mineral oil, its insulating and cooling fluid, and various electrical tests of its paper and oil insulation system. One of the most important indicators, in this writer's opinion, is testing the transformer's mineral oil for dissolved combustible gases. This is analogous to blood testing of humans to detect various biological abnormalities.

The significance and use of dissolved gas as a diagnostic tool to determine the serviceability of a transformer will be discussed in this paper. ANSI/IEEE C57-104 1978, "IEEE Guide for the Detection of Generated Gases in Oil-Immersed Transformers and their Relation to the Serviceability of the Equipment," hereafter referred to as C57.104, will be reviewed. C57.104 is undergoing a major revision in the area of interpretation and the decision process.

It is this aspect of C57.104 that will be addressed from the viewpoint of the user.

This paper is organized in the sequence proposed for the revised C57.104 and is the basis of the decision process.

1. Detection.
2. Diagnosis.
3. Evaluation and action.

Detection of a Problem

As the mineral oil circulates through the transformer, it carrys with it solids and gases. Using the human body and its blood circulation system as an analogy, a doctor can detect the early warning signs of many abnormalities from tests of blood samples. It is likewise possible to detect the early signs of abnormalities in a transformer by sampling its vital fluid (oil) and analyzing it for combustible gas and other indicators.

There are several monitoring options available to the user which can provide an indication of the presence of combustible gas. Some are a gas detector relay on conservator and membrane

---

[1]Consulting engineer, Consolidated Edison Co. of New York, New York, NY 10003.

sealed units, total combustible gas on gas-blanketed units, dissolved gas in oil using periodic manual sampling, or continuous monitoring using installed hydrogen detection devices. These may be used alone or in combination. The existing C57.104 indicates normal operating levels for total combustible gas and gas in oil. In my company we manually sample transformer oil on a routine schedule as indicated below for chromatographic analysis:

Generator step-up transformers	Every two months
345-kV transformers	Quarterly
138 and 69 kV	Annually

We also monitor several units using fixed detection devices, which indicates hydrogen content dissolved in the oil.

It is recognized that all transformers will produce combustible gases as a result of the normal aging process. That being true, it should then be easy to determine abnormal transformer behavior by comparing it to the normal aging gassing quantities. That would be fine if one knew what gases and levels are associated with a healthy transformer. Since this is still an art and not a science, we must depend on our documented experience, which is small compared to that in Europe.

In order to help determine norms for a healthy transformer, we must rely on published information like that presented in C57.104 and other sources as indicated in Table 1. It is evident from data in Table 1 that there is no agreement on acceptability levels.

Diagnosis of the Gas Data

Let us assume that we have determined that a transformer has abnormal gas content. What next? At our company we resample the unit's oil to verify the gas content. If it is still high we monitor the unit by sampling the oil every week or daily if the total gas content is high or if acetylene is above 25 ppm. We observe the gassing rate, and if it is essentially stable we will only monitor; however, if it is increasing more than 100 ppm per day we will take it out of service for an internal inspection.

Before removing the unit from service, it would be helpful if one could diagnose the source of the gassing, much like making a diagnosis of a human illness based on blood tests. There are several available methods used to diagnose the source of gassing. Three of the most commonly used methods are:

1. Dornenberg ratio method.
2. Rogers ratio method.
3. Key gas method.

TABLE 1—*Dissolved gas-in-oil acceptability limits from various sources.*

Source	H_2	CO	CH_4	C_2H_6	C_2H_4	C_2H_2	TCG
Electra (1978) [2]	28.6	289	42.2	85.6	74.6	. . .	520
Hydro-Quebec [3]	500	900	35	20	85	2	1600
General Electric [4] (3-year-old Trf)	200	200	50	20	100	20	595
General Electric [4] (6-year-old Trf)	500	500	100	400	200	25	1725
Doble [2]	100	250	100	60	100	5	<700
IEEE [5]							
Generator	240	580	160	115	190	11	1296
Transmission	100	350	120	65	30	35	700

NOTE: TCG = total combustible gas; trf = transformer.

TABLE 2—*Transformer gassing history.*

Gas	Sampling Dates		
	1/5/87, ppm	4/7/87, ppm	4/10/87, ppm
Hydrogen (H$_2$)	ND	723	2 036
Methane (CH$_4$)	48	1988	4 743
Carbon monoxide (CO)	14	28	28
Carbon dioxide (CO$_2$)	412	496	421
Ethane (C$_2$H$_6$)	53	595	1 303
Ethylene (C$_2$H$_4$)	5	3259	8 284
Acetylene (C$_2$H$_2$)	ND	32	110
Total combined gas	120	6625	16 504
Rate/day (ppm/day)	. . .		3 293

NOTE: The transformer above is a 138-kV, 200-MVA Phase Angle Regulator. The volume of oil was about 20 000 gal.

Dornenberg Method

This method uses the ratio analysis of significant gases, that is, methane/hydrogen, acetylene/ethylene, acetylene/methane, and ethane/acetylene. Dornenberg requires, for accuracy, that the significant gases exceed norms developed from the analysis gas samples from a large population of transformers.

Rogers Method

This method also uses ratios of significant gases, except that Rogers uses three ratios, that is, methane/hydrogen, acetylene/ethylene, and ethylene/ethane. Rogers does not indicate any qualifying norms when using his method.

Key Gas Method

This method uses the percentages of various key gases to determine the cause, that is, in the case of overheating oil the key gas is ethylene with smaller quantities of methane, ethane, and hydrogen present. The key gas in the case of arcing is acetylene, and in the case of corona the key gas would be hydrogen.

However, unlike the diagnosis of blood where rules have been established, the diagnosis of gas in oil is not so clearly defined. It has been reported that the diagnosis of gases do not always yield the same results when using different methods on the same sample [1].

The presence and quantity of gases are dependent on many variables, such as type, area, and temperature of the fault; solubility and degree of saturation of various gases in oil; the type of oil preservation system; the type and rate of oil circulation; the kinds of material in contact with the fault; and, finally, variables associated with the sampling and measuring procedures themselves. Keeping in mind that this is an art and not a science, it is safe to use these methods as indicators.

One of the obstacles to better accuracy is the lack of a large published data base in the United States which relates gassing to actual conditions found. The Edison Electric Institute (EEI) is beginning to collect data on gassing transformers where the cause of gassing has been determined.

Evaluation and Action

For illustration purposes, let us assume that we have a transformer that has developed a gassing history (see Table 2 for gas data). Using Fig. 1, a flowchart, I will address the evaluation process being considered in the revision of C57.104.

1. Gas-in-oil monitoring indicates abnormal levels of hydrogen, methane, ethane, ethylene, and acetylene based on chromatographic analysis.
2. Unit is resampled and the high gas levels are confirmed.
3. Gas level increasing.
4. Notified manufacturer.
5. Diagnosis made using the following methods with results as indicated in Fig. 2.
6. Continued to monitor unit and arranged for outage to inspect.
7. Inspection revealed that core laminations were touching the tank wall (shell form design).
8. Enter information in company and EEI data base.

Conclusion

The new C57.104 will be more informative and therefore more useful to the user. However, as was pointed out, we need a broader data base of United States utility and industrial experience with gassing transformers. This is necessary in order to improve the accuracy of diagnosis, which is a key element in the decision-making process. No one wants to remove equipment from

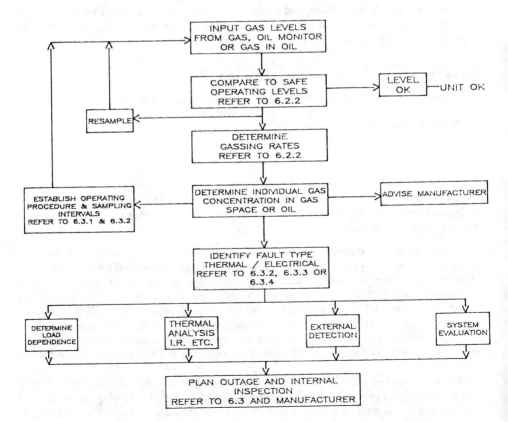

FIG. 1—*Flowchart. References refer to the section of the proposed C57.104.*

DORNENBERG METHOD

$\dfrac{CH4}{H2}$	$\dfrac{C2H2}{C2H4}$	$\dfrac{C2H6}{C2H2}$	$\dfrac{C2H2}{CH4}$

Using the data from 4/7/87 in Table 2:

$$\dfrac{1988}{723} = 2.7 \quad \dfrac{32}{3259} = 0.01 \quad \dfrac{595}{32} = 18.6 \quad \dfrac{32}{1988} = 0.02$$

Diagnosis: Hot spot - No cellulose

ROGERS METHOD

$\dfrac{CH4}{H2}$	$\dfrac{C2H6}{CH4}$	$\dfrac{C2H4}{C2H6}$	$\dfrac{C2H2}{C2H4}$

Using the data from 4/7/87 in Table 2:

$$\dfrac{1988}{723} = 2.7 \quad \dfrac{595}{1988} = 0.3 \quad \dfrac{3259}{595} = 5.5 \quad \dfrac{32}{3259} = 0.01$$

Diagnosis: Circulating current core and tank.

KEY GAS METHOD

Using data from 4/7/87 in Table 2:

GAS	PERCENT OF TOTAL
H2	11
CH4	30
C2H6	9.6
C2H4	49
C2H2	0.4

Diagnosis: Overheated oil

Condition found when inspected: Core steel touching tank wall.

All methods were in basic agreement.

FIG. 2—*Example of diagnostic methods.*

service without technical justification. I'm sure that data collection is not the whole answer, but it is a step in the right direction and can eventually lead to expert systems that are dependable.

References

[*1*] Sobral, C. L. C., "Correlation Between Results of Dissolved Gas Analysis and Actual Transformer Inspection," Doble Conference Paper, 1986, Sec. 6-501, The Doble Conference Minutes, Doble, Inc., Boston, MA.

[2] Baker, A. E., Griffin, P. J., and Locke, C., "An Update on Fault-Gas Analysis (A Review)," Doble Conference Paper, 1986, Sec. 10-301, The Doble Conference Minutes, Doble, Inc., Boston, MA.

[3] Dind, J. E. and Regis, J., "Preventative Electric Maintenance," *Pulp and Paper Canada,* Vol. 76, No. 9, September 1975, pp. 61–64.

[4] General Electric Bulletin GET-6552, "Power Transformer Dissolved Gas Analysis," General Electric, Pittsfield, MA.

[5] ANSI/IEEE C57-107 1978, "IEEE Guide for the Detection and Determination of Generated Gases in Oil-Immersed Transformers and their Relation to the Serviceability of the Equipment," under revision.

Paul J. Griffin[1]

Criteria for the Interpretation of Data for Dissolved Gases in Oil from Transformers (A Review)

REFERENCE: Griffin, P. J., **"Criteria for the Interpretation of Data for Dissolved Gases in Oil from Transformers (A Review),"** *Electrical Insulating Oils, STP 998,* H. G. Erdman, Ed., American Society for Testing and Materials, Philadelphia, 1988, pp. 89–106.

ABSTRACT: The analysis of gases dissolved in oil has been used as a diagnostic tool for many years to determine the condition of transformers. The criteria used in evaluating dissolved gas-in-oil data are reviewed. The criteria are based on experience from failed transformers, transformers with incipient faults, laboratory simulations, and statistical studies. The interpretation of results is enhanced by including specific information on a particular transformer, including its history. Comparison of a particular transformer of concern with a population of similar transformers provides additional information as regards "normal" gassing behavior. The interpretation of dissolved gas-in-oil data is not a simple matter and requires the integration of numerous criteria. Diagnostic gases for silicone fluids are also discussed.

KEY WORDS: mineral oil, transformer oil, silicone fluid, dissolved gases, fault gases, combustible gases, gas solubility, key gases, ratios, overheating, corona, arcing

It has long been recognized that the degradation of transformer insulation results in the formation of by-products, notably characteristic gases. In a sense, the generation of these gases, quantified by routine methods, is rather fortuitous since the detection and measurement of abnormal amounts provides a basis for appropriate measures to be undertaken to prevent unplanned outages and/or catastrophic failures. Although the utility industry uses the dissolved gas-in-oil test as a routine tool, there is no universally accepted means for interpreting results. There have been many refinements of existing methods of interpretation; however, no single "simple" method seems to provide a complete picture under all circumstances. A simple "litmus paper" type of approach with black and white answers is not sufficient for resolving problems of this complexity. Although general rules can be used, as much specific information as possible is important in providing the diagnosis on a particular transformer. This paper reviews some of the important criteria in the interpretation of results of tests for dissolved gases in oil from transformers.

Historical Information

Dissolved gas-in-oil data by themselves do not always provide sufficient information on which to evaluate the integrity of a transformer system. The nameplate information and the history of the transformer in terms of maintenance, loading practice, previous faults, etc. are an integral part of the information required to make an evaluation. For example:

1. How old is the transformer?
2. Did a bushing fail at some point?
3. Did the transformer fail at some point?

[1]Laboratory manager, Doble Engineering Co., Watertown, MA 02172.

4. Is the unit heavily loaded or did the unit have to carry overload for a period of time?
5. Has the unit been repaired after a failure (residual gases)?
6. Have previous tests been performed to determine gas-in-oil results?
7. Has the percent total combustible gas (TCG) in the gas space risen suddenly?
8. Has the percent TCG in the oil risen suddenly?
9. Has the oil been degassed?

As one can see, the answers to these questions may very well temper the evaluation of a particular transformer.

Oil Preservation Systems

It has been common practice since the 1950s to design transformers such that the oxygen (air) concentration can be reduced and maintained at low levels to preserve the oil and cellulose insulation. Prior to the advent of oil preservation systems, transformers were designed to freely "breathe out" gases generated and to "breathe in" fresh air (oxygen). The oil preservation systems are of three basic types as seen in Fig. 1:

1. Gas blanketed (usually nitrogen).
2. Conservator (sealed with a bladder).
3. Conservator (opened to atmosphere).

The gas-blanketed transformer can be either a sealed unit or a pressurized system containing a replenishing bottle. Overpressurization is prevented through the use of a pressure release valve. The gas blanket type of system provides a cushion for expansion and contraction of the oil. Gases partition between the gas blanket and oil in accordance to Henry's law, which essentially says that the concentration of gas in the gas phase is directly proportional to its concentration in the liquid phase. The Ostwald coefficient, the volume of gas dissolved in a unit volume of oil at equilibrium at a specified partial pressure and temperature, is used to calculate the concentration of gas in one phase when the concentration in the other phase is known. Table 1 gives the Ostwald coefficients for an oil with a density of 0.880, a temperature of 25°C, and pressure of 1 atm for nine characteristic gases [1].

When the concentration of gas in the oil is known, an estimate of the percent concentration of each gas in the gas blanket is given by:

$$\frac{\dfrac{\text{ppm gas in oil}}{\text{Ostwald coefficient}}}{\Sigma \dfrac{\text{ppm gas in oil}}{\text{Ostwald coefficient}}} \times 100\% = \% \text{ each gas in gas space}$$

The gases H_2, N_2, O_2, and CO have relatively low solubilities in oil, with CH_4 somewhat higher. The gases CO_2, C_2H_2, C_2H_4, and C_2H_6 have high solubilities in oil. An example of how this affects the partitioning of gases between the two phases is given in Table 2.

The majority of the gas is typically nitrogen; however, for demonstration purposes an approximate concentration of O_2, as would be found in air, is included. As can be seen, the O_2/N_2 ratio increases in the oil due to the higher solubility of O_2 in oil. Of the combustible gases, H_2 and CO preferentially partition into the gas space and, therefore, are often the major components. Hydrocarbons, most importantly acetylene, tend to remain in the oil. These estimated values are based on certain equilibrium conditions and therefore are only rough guidelines.

The solubility of gases in oil varies with temperature and pressure. H_2, N_2, CO, and O_2 increase in solubility with temperature, and CO_2, C_2H_2, C_2H_4, and C_2H_6 decrease in solubility with temperature, while CH_4 remains essentially unchanged [2–4]. All the gases are proportion-

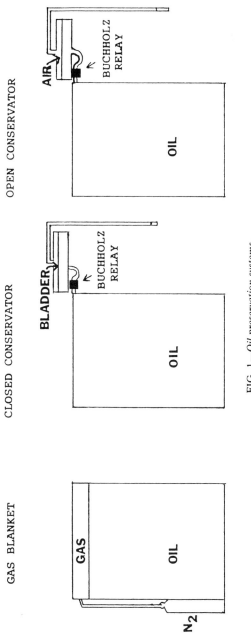

FIG. 1—*Oil preservation systems.*

TABLE 1—*Ostwald solubility coefficients.*

Gas	Ostwald Coefficient
O_2	0.138
N_2	0.0745
CO_2	0.900
CO	0.102
H_2	0.0429
CH_4	0.337
C_2H_6	1.99
C_2H_4	1.35
C_2H_2	0.938

ally more soluble in oil with increasing pressure [*3*]. For example, doubling the pressure doubles the gas concentration in the oil. The solubility of the gases also modestly increases as the density of the oil decreases [*1,3*].

In conservator systems there is no gas blanket in the transformer main tank and the entire transformer is filled with oil. An expansion chamber permits oil volume changes due to temperature. The expansion chamber can be closed by a bladder type of arrangement which permits changes in oil volume while restricting contact with the air, or can be open in contact with the air. The closed system essentially results in complete solution of the gases into the oil except in the case where gas bubbles are formed and collect in a special device, such as a Buchholz relay, for such conditions. A Buchholz relay (positioned high on the transformer between the main tank and conservator tank) collects gas that comes out of solution and allows a floating contact(s) to move. If the float moves sufficiently it closes a contact and causes an alarm. A second contact can initiate breaker tripping if sufficient gas is present. The open type conservator system allows for a continuous exchange of gases at the air/oil interface of the expansion chamber; however, access is somewhat restricted. In the case of the completely closed conservator system, hydrogen, carbon monoxide, and to a lesser extent methane will be proportionally higher than in the gas-blanketed-type system. In the open system these same gases will preferentially decrease in concentration as they are slowly lost to the atmosphere through the expansion chamber.

TABLE 2—*Estimation of gas composition in the gas space.*

Gas	Gas in Oil, ppm (Vol./Vol.)	Estimated ppm Gas in Gas Space[b]	Estimated % Gas in Gas Space
O_2	28 400	206 000	20.6
N_2	59 000	792 000	79.2
CO_2	1 000	1 110	0.11
CO^a	100	980	0.10
$H_2{}^a$	100	2 330	0.23
$CH_4{}^a$	100	297	0.03
$C_2H_6{}^a$	100	50	0.01
$C_2H_4{}^a$	100	74	0.01
$C_2H_2{}^a$	100	107	0.01

[a]Combustible gases.
[b]Estimated value under equilibrium conditions at 25°C and 1 atm.

Homogeneity of Gases in Oil

The homogeneity of the gas in the oil is dependent upon factors such as rate of gas generation, access of the fault area to flowing oil, rate of oil mixing, presence of a gas blanket, and, to a small degree, diffusion. In the vast majority of the cases, incipient fault gases are generated in a specific area in the oil phase and then transferred throughout the system. If gas bubbles are generated, the gases will immediately begin to partition between the gas bubble and the oil according to their solubilities, with the more oil-soluble gases going in the oil. Hydro-Quebec Institute of Research has developed a means for estimating the depth of a fault in oil by comparing the proportions of H_2, CH_4, and C_2H_2 present in the gas in the Buchholz relay and in the oil immediately following an alarm [5]. The further the distance the bubble travels in the oil, the greater the differences in the relative composition of the gases in the relay and the oil. Sometimes an incipient fault can also be further located by comparing combustible gas levels for oil samples taken at various levels in the transformer, that is, the greater concentration of gases is nearest the source of the fault.

Gas Bubbles

The formation of gas bubbles in power transformer systems is a serious and complex problem. Gas bubbles can impair the dielectric integrity of the transformer system with severe consequences. Three mechanisms have been postulated concerning their formation [6]. These are:

1. Supersaturation of the oil with gas.
2. Thermal decomposition of the cellulose insulation.
3. Vaporization of adsorbed moisture in the cellulose.

In a pressurized transformer which is rapidly cooled after being at a high temperature, there is a large pressure drop due to the reduction in oil volume. The lower pressure and temperature results in excess gas being present, and saturation or supersaturation of the oil with the blanketing gas occurs. Consequently, gas bubbles may be released if initiated by mechanical vibration, oil pump operation, or sufficient electrical stress [7-9]. The other known causes of bubble formation, generation of CO and CO_2 from the degradation of the cellulose insulation, and vaporization of moisture, are subject to higher temperatures associated with transformer overloading (140°C winding hot-spot temperature) [6,10,11]. It also appears that the rate of temperature rise and the percentage of moisture in the paper insulation may be important factors [11].

Nitrogen and Oxygen

Nitrogen and oxygen are not typically considered diagnostic gases per se, but nevertheless are useful to assess the presence of leaks, overpressurization, or changes in pressure and temperature on the transformer system.

Oxygen is one of the key promoters of oil and cellulose oxidation in transformer systems and, therefore, is generally maintained at low concentrations. Increases in oxygen suggest a leak in the system. However, a determination of high oxygen may also be the result of a leak during sampling and testing of an oil and, therefore, should be confirmed. A rapid decrease in oxygen with changes in oil properties and the generation of other gases may be indicative of overheating.

Changes in the nitrogen content can be due to changes in the temperature of the transformer in gas-blanketed systems. As the insulating oil heats and cools, there is a corresponding increase and decrease in the oil volume and compression and relaxation of the gas space. The increase and decrease in pressure result in raising and lowering the dissolved N_2 (and other gases) content in the oil. Excessive N_2 can lead to bubble formation, as previously discussed.

Combustible Gases and Carbon Oxides

Combustible gases and carbon oxides are the characteristic diagnostic gases generated from incipient fault conditions. As expected, the normal aging of transformers also results in the generation of these gases. Excessive generation of these gases due to abnormal conditions or accelerated deterioration of the insulating materials is determined by comparing individual test results to empirical norms for a population of transformers, total combustible gases, key gases, ratios, trends, and individual "fingerprints."

Norms

A means of establishing if a transformer with no measured previous dissolved gas history is behaving normally is to compare it with the gassing characteristics exhibited by the majority of similar transformers or a "normal population." As the transformer ages and gases are generated, the normal levels for 90% of a typical transformer population can be determined. From these values, and based on experience, acceptable limits have been determined such as those given in Table 3 [12].

Considerable differences are apparent for what is considered a normal transformer with acceptable concentrations of gases. Some of the differences seen here may be due to differences in the transformer populations studied; that is, gas blanketed versus conservator, generation versus transmission, age of transformer, etc. Although not listed here, H_2 and C_2H_2 limits are based on the main tank and load tap changer having separate compartments. If the oil is common to both compartments, then revised H_2 and C_2H_2 values are required.

Norms are now being developed for factory tests on new power transformers. Typical results from three sources for factory tests for dissolved gas in oil performed after the heat run are given in Table 4 [13,14].

Heat run data are not always reported for the same time interval, and, therefore, comparison of the different levels here is not possible; however, in general, results are typically low on a new transformer.

By definition most transformers, when tested for gases dissolved in oil for the first time, will be "normal" when compared to any of the acceptance limit values. When an abnormal situation is indicated, a testing schedule is devised with increased sampling frequency to determine a trend. Only in very rare severe cases would an investigation be required based on a first-time analysis.

TABLE 3—*Dissolved gas acceptable limits—various sources.*

	Gas, ppm (Vol./Vol.)							
	H_2	CO	CH_4	C_2H_6	C_2H_4	C_2H_2	CO_2	TCG
Doble	100	250	100	60	100	5	...	610
IEEE[a]								
Generator	140	580	160	115	190	11	...	1296
Transmission	100	350	120	65	30	35	...	700
Electra (CIGRE)[b]	28.6	289	42.2	85.6	74.6	...	3771	520
Manufacturer	200	500	100	100	150	15	...	1065
	(250)	(1000)	(200)	(200)	(300)	(35)	...	1985

[a]In the process of being revised.
[b]Corrected values 1978.
( ) = Values 6 to 7 years.

TABLE 4—*Factory dissolved gas-in-oil tests after heat runs, typical values.*[a]

Gas	1	2	3[b]
H_2	5	5	30
O_2	. . .	. . .	. . .
N_2	. . .	. . .	. . .
CO	10	75	75
CO_2	100	400	400
CH_4	5	5	5
C_2H_6	5	5	5
C_2H_4	5	5	5
C_2H_2	0	None	None detected

[a]Results reported in ppm (vol./vol.).
[b]Upper limits for 90% of units tested for 24-h heat run.

Total Combustible Gas (TCG) Limits

On a first-time sample, the severity of an incipient fault can be further evaluated by the TCG present. Limits for TCG, modified since originally published in 1973 [*15*], are as follows:

Combustible Gas Limits, ppm

0–500   Unless individual gas-acceptance values are exceeded, TCG in this range indicates the transformer is operating satisfactorily.

501–1500   Indicates some decomposition of the transformer insulation has occurred and action should be taken to establish the trend to determine if an incipient fault is likely to evolve as a problem.

1501–2500   A high level of decomposition of the transformer insulation has occurred and a fault may be present. Immediate action should be taken to establish the trend to determine if the fault is progressively becoming worse.

>2500   Substantial decomposition may have occurred to the integrity of the transformer insulation and therefore could result in a failure. The gassing rate and the cause of gassing should be identified and appropriate corrective action should be taken.

Key Gases

Combined, the individual gas acceptance values and total combustible gas, while informative, do not provide sufficient information to assess a transformer's condition, even for a first-time analysis. A further requirement is the identification of the type of fault. Characteristic "key gases" have been used to identify particular fault types. Laboratory simulations, and comparison of results of dissolved gas-in-oil tests combined with observations from the teardown of failed transformers, have permitted the development of a diagnostic scheme of the characteristic gases generated from the thermal and electrical (corona and arcing) deterioration of electrical insulation [*16*]. Figure 2 lists the gases for the conditions of arcing, corona, overheating in oil, and overheating in paper in the order of decreasing severity.

For each type of condition, the level of concern at each total combustible gas level would vary. In the case of electrical problems, erratic and rapid deterioration of the insulation is more likeiy, and (particularly in the case of arcing) either high or very low levels of gases can be generated before failure occurs. In general, the overheating condition is considered the least serious because, in most cases, deterioration of the insulation is slow and relatively high levels of combustible gases are generated as by-products. This permits ample time to plan and carry out

1. ARCING

Large amounts of hydrogen and acetylene are produced, with minor quantities of methane and ethylene. Carbon dioxide and carbon monoxide may also be formed if the fault involves cellulose. The oil may be carbonized.

Key gas—Acetylene

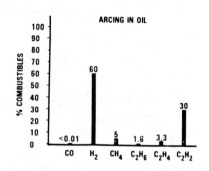

2. CORONA

Low-energy electrical discharges produce hydrogen and methane, with small quantities of ethane and ethylene. Comparable amounts of carbon monoxide and dioxide may result from discharges in cellulose.

Key gas—Hydrogen

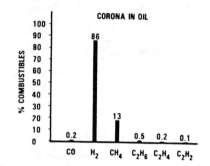

3. OVERHEATED OIL

Decomposition products include ethylene and methane, together with smaller quantities of hydrogen and ethane. Traces of acetylene may be formed if the fault is severe or involves electrical contacts.

Key gas—Ethylene

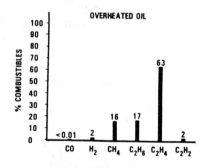

4. OVERHEATED CELLULOSE

Large quantities of carbon dioxide and carbon monoxide are evolved from overheated cellulose. Hydrocarbon gases, such as methane and ethylene, will be formed if the fault involves an oil-impregnated structure.

Key gas—Carbon Monoxide

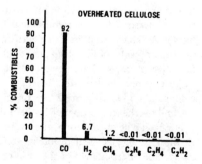

FIG. 2—*Key gases.*

remedial action. It should be noted that the relative concentrations of the gases given in Fig. 2 are only guidelines and variances do occur. In Table 5 there are three examples showing the difference in the concentration of characteristic gases for each condition which resulted in remedial action by the owners of the transformers [17–19].

Transformer 1 involved a case where low energy arcing was occurring [17]. Levels of combustible gases were stable for about 13 months at the level shown in Table 5, except for hydrogen, which showed a level of 150 ppm in one test. Test results on a sample seven months later again showed an increased level of H_2 (90 ppm) and an increasing level of C_2H_2 (78 ppm). The unit was removed from service and investigated. A piece of sheet steel which was found in an insulating phase barrier was apparently arcing to the transformer frame. Serious damage could have occurred if this problem had not been identified and corrected.

Transformer 2 showed an increase in H_2 in the oil from 1700 to 2300 ppm over a period of seven months before it was removed from service and repaired for loose connections [18].

Transformer 3 had an overheating condition due to a recurring core ground problem which was eventually corrected by burning it away [19].

Several refinements to the key gas method of identifying incipient faults can be seen for arcing and overheating conditions. When arcing of low energy or of short duration occurs, there will be less overheating of the oil and, hence, less C_2H_4, CH_4, and C_2H_6 formed. Even though the arc is of a sufficient hot-spot temperature to yield C_2H_2, cooler transition zones about the arc will generate C_2H_4, CH_4, and C_2H_6 if sufficient time or energy is available to effect a large portion of the oil.

Overheating of oil can be further characterized based on thermodynamic considerations [20]. As temperature increases, the concentration of gases increases as follows:

Low Temperature ($\cong 120°C$)	CH_4
	C_2H_6
	C_2H_4
High Temperature ($> 700°C$)	C_2H_2

At specific temperatures, one or two of these gases predominate while lesser quantities of the other gases are formed. In a practical situation, two scenarios are likely to occur which obscure this clear temperature discrimination. In one case, a thermal condition deteriorates over time,

TABLE 5—*Transformers with incipient faults.*

Gas, ppm (Vol./Vol.)	Transformer		
	1	2	3
H_2	0	1 700	540
O_2	1 100	3 000	2 300
N_2	79 000	110 000	87 000
CH_4	9	43	1 300
CO	33	440	420
C_2H_6	7	6	160
CO_2	510	8 400	2 000
C_2H_4	8	2	810
C_2H_2	9	0	2
TGC[a]	80 676	123 591	94 532
TCG[b]	57	2 191	3 232

[a]Total gas content.
[b]Total combustible gas.

resulting in a changing spectrum of gases as the temperature increases. In the other case, a large thermal gradient exists, giving rise to a wide spectrum of gases.

It should also be noted that other factors, such as high O_2, can result in additional gases being generated, that is, CO and CO_2.

Table 6 shows four cases of transformers exhibiting evidence of overheating in oil by the characteristic hydrocarbon gases (in decreasing order of severity).

In Case 1, the major gases are CH_4, indicating low temperature overheating in oil, and C_2H_4 along with traces of C_2H_2, indicating very high temperature. This transformer had a core ground problem which generated gases at a rapid rate, suggesting that high temperature overheating (hot spot) was occurring and a thermal gradient existed with a large area of more low-to-moderate overheating also occurring.

In Case 2, the predominant gas is C_2H_6. This is unusual in that in most cases where C_2H_6 or CH_4 is the predominant gas (with C_2H_4 a small proportion) modest gassing rates are evident. Gas generation due to heating follows Arrhenius-type kinetics [16] and, therefore, at low temperatures the characteristic gases should be generated in relatively small quantities, although varying with the size of the incipient fault area. In Case 2 there are a number of transformers from this same manufacturer which show the same condition which appears to stabilize after an initial "break-in" period. The gas content in this unit has not changed in three years. Cases 3 and 4 indicate lower temperature overheating (predominate hydrocarbon gas is C_2H_6 and CH_4, respectively) and, as would be typical, lower total gaseous hydrocarbons.

Ratios

Several analytical schemes have been developed which employ and compare ratios of characteristic gases generated under incipient fault conditions. The methods are used to determine the type of incipient fault condition similar to the key gas method. The advantages to the ratio methods are that they are quantitative, independent of transformer oil volume, and can be computer programmed. The disadvantages are that they may not always yield an analysis or may yield an incorrect one and, therefore, should only be used as an additional guideline to other diagnostic methods such as key gas methods. In addition, ratio methods require minimum levels of gases to be present before they are reliable and better precision on the part of the analysts performing the dissolved gas-in-oil test. Several papers have discussed problems with ratio methods [21–22].

TABLE 6—*Transformers exhibiting overheating of oil.*

Gas, ppm (Vol./Vol.)	Transformer			
	1	2	3	4
H_2	540	1	16	110
O_2	2 300	7 200	3 700	3 100
N_2	87 000	110 000	100 000	82 000
CH_4	1 300	69	290	110
CO	420	400	240	140
C_2H_6	160	2 300	480	39
CO_2	2 000	6 800	4 400	1 500
C_2H_4	810	180	33	8
C_2H_2	2	0	0	0
TGC[a]	94 532	126 950	109 159	87 007
TCG[b]	3 232	2 950	1 059	407

[a]Total gas content.
[b]Total combustible gas.

Two ratio methods most commonly used are those of Dornenburg [23] and Rogers [24–26]. In both methods the same gases are used, although some of the ratios are different, as can be seen in Figure 3. In both cases involvement of solid insulation (CO, CO_2) is handled separately.

These methods and ratio methods in general are to some extent dependent upon the distribution of the gases in the transformer system; that is, there are differences for conservator and gas-blanketed units.

Both Dornenburg's and Rogers' ratio methods were devised from data from conservator-type transformers with expansion chambers exposed to the atmosphere. This type of transformer experiences a slow loss of the less soluble gases (due to restricted oil movement between the main tank and the expansion chamber) at the oil/air interface. Oommen has determined correction factors for the ratios for conversion between sealed conservator and gas-blanketed transformers based on the solubility of the gases in oil [3] (calculation of the rate of loss of gases from the open expansion tank conservator is not a constant and, therefore, difficult to calculate). The correction factor is usually only significant for the less soluble gases, that is, CH_4/H_2 ratio. Dornenburg developed ratios for gases dissolved in oil and gases in the Buchholz relay with the relation between the two based on the solubility of each gas in oil.

Although the Dornenburg's and Rogers' ratios yield similar interpretation of dissolved gas-in-oil data, there is a difference in the complexity of the schemes. The Dornenburg method, used when prescribed normal levels of gassing are exceeded, provides a simple scheme for distinguishing between overheating, partial discharge, and other types of discharge.

The Rogers' ratio method is a more comprehensive scheme which details temperature ranges for overheating conditions (based on Halstead's research [20]) and some distinction of the severity of incipient electrical fault conditions. A normal condition is also listed. Minimum gas limits for using the Rogers method are very low (10 ppm H_2, 2 ppm hydrocarbons [24]). It is this author's opinion that the use of low minimum limits can result in erroneous conclusions and that limits such as those used by Dornenburg are more appropriate.

Several examples may illustrate some of the advantages and disadvantages of using ratio methods. If we examine the data given in Table 6 for the various overheating conditions, we arrive at the following diagnosis, according to Rogers' code [24,26]:

Transformer 1: Core and tank circulating currents, overheated joints.
Transformer 2: Overheating 150 to 200°C.

ROGERS	DORNENBURG
$\dfrac{CH_4}{H_2}$	$\dfrac{CH_4}{H_2}$
$\dfrac{C_2H_6}{CH_4}$	$\dfrac{C_2H_6}{C_2H_2}$
$\dfrac{C_2H_4}{C_2H_6}$	$\dfrac{C_2H_2}{CH_4}$
$\dfrac{C_2H_2}{C_2H_4}$	$\dfrac{C_2H_2}{C_2H_4}$

FIG. 3—*Ratio methods.*

Transformer 3: Overheating 150 to 200°C.

Transformer 4: Slight overheating—below 150°C.

As can be seen, this offers a refinement to the key gas method.

In contrast, Table 7 shows two cases published by Vieira [27] where both Rogers' and Dornenburg's ratio methods did not provide a diagnosis, even though it is obvious a very serious incipient fault condition existed.

Trend Analysis

When a possible incipient fault condition is identified by the first time analysis, the gassing trend should be determined. Subsequent analysis provides information as to which gases are currently being generated and the rate of generation of these gases. Even when subsequent analyses are of a routine nature, they provide the baseline from which to judge future occurrences. In the examination of trends, key gases, total combustible gas, rate of gas generation, and "fingerprints" (of normal trends) of particular types of transformers are considered.

As in the case of a first-time sample, key gases are used to determine the type of incipient fault condition with subsequent samples, with the difference being that only those gases actively being generated are used in the diagnosis of subsequent samples. If no significant change in gas concentration is observed for a reasonable period of time, even though values exceed norms, or ratios indicate that a problem exists, no problem may be indicated unless excessive insulation has deteriorated (which usually results in more gassing). Ratio methods may be used to aid in the analysis, but, depending upon the history of the transformer, they may not be able to distinguish between past and present conditions. For example, a transformer which has been degassed after a fault will show an altered gas pattern as the less-soluble gases are preferentially removed. Also, test results for a transformer with more than one type of fault may be difficult to interpret using ratio methods. Again, the total combustible gas content (in oil) is used to judge the relative deterioration of the insulation. It is possible that an apparent incipient fault condition will generate gases away from insulating structures and cause no harm to the transformer insulation system. However, such a condition may also mask the effects of a more serious condition developing at the same time.

TABLE 7—*Gas-in-oil test results.*

Gas, ppm (Vol./Vol.)	Case 3	Case 7
H_2	650	76
O_2	2 900	3 500
N_2	53 000	93 000
CH_4	81	6 000
CO	380	36
C_2H_6	170	27 000
CO_2	2 000	4 400
C_2H_4	51	120 000
C_2H_2	270	1 700
TGC[a]	59 502	255 712
TCG[b]	1 602	154 812

NOTE: *Actual Inspection.* Case 3: Electrical problem caused by poor shield contact. Case 7: Metal particles, carbon located on core; burning and erosion by electric current noted.

[a]Total gas content.

[b]Total combustible gas.

The rate of gas generation which can be load-dependent provides information as to the severity of the condition and offers guidelines as to whether or not a unit should be removed from service. Two means for assessing the gassing rate are to use the change in concentration of gas in parts per million or to determine the actual amount of gas generated. General guidelines that have been used for removing transformers from service are 100 ppm/day [14] and 0.1 ft^3 (0.003 m^3) gas per day [28,29]. Cubic feet can be converted to ppm using:

$$ppm = \frac{(gas\ in\ ft^3)(7.48)(10^6)}{gallons\ of\ oil}$$

If results are in cubic metres they can be converted to cubic feet by multiplying cubic metres times 35.3. Thus, in 10 000 gallons of oil, 0.1 ft^3 (0.003 m^3) gas per day is equivalent to 75 ppm/day. When the nature of an incipient fault is known, the use of one of the limits may be preferred. It may be more appropriate to use the absolute rate of gas generation for situations involving localized conditions (where dilution of the gas into different volumes of oil would alter relative ppm values).

Conversely, the ppm limits may be more appropriate for use when nonlocalized conditions are involved, such as general conductor overheating. A further refinement by McNutt et al. [11] in determination of cellulose degradation takes into account surface area and temperature involved. These guidelines for gassing rates are general "rules of thumb" and considerations such as type of fault, previous history of the particular type of transformer, etc. must always be taken into account.

Fingerprints

Transformers from the same manufacturer and of the same type exhibit what might be called "normal trends," "characteristic trends," or "fingerprints" [12]. As a new transformer "breaks in," some gas evolution is expected and may occur at an initially high rate which subsequently slows down or plateaus. Specific manufacturer's transformer designs may show levels and types of gases which are characteristic and do not represent an incipient fault condition.

Tables 8A, 8B, and 8C show what may be considered "fingerprints" of three different transformer designs. In each case two examples with initial test values and those after two or three years are given.

Table 8A gives two examples of typical values from a data base of 50 to 100 transformers which can be considered as representative of the group.

It is obvious that one can single out the predominant "key" gases, hydrogen and carbon monoxide, and by cursory inspection one might even conclude that a possible problem exists.

TABLE 8A—*Transformer fingerprints.*

Gas, ppm (Vol./Vol.)	Example 1		Example 2	
	I	3	I	3
H_2	350	260	110	210
CH_4	44	61	11	13
CO	670	650	520	630
C_2H_6	26	25	3	4
CO_2	3000	1900	5000	3900
C_2H_4	9	5	8	10
C_2H_2	...	...	...	...

NOTE: I = initial.

TABLE 8B—*Transformer fingerprints.*

Gas, ppm (Vol./Vol.)	Example 1		Example 2	
	I	2	I	2
H_2	20	12	0	0
CH_4	55	16	60	29
CO	2	9	0	0
C_2H_6	250	83	61	46
CO_2	1900	1500	650	250
C_2H_4	500	620	200	150
C_2H_2	1	Trace	0	0

NOTE: I = initial.

TABLE 8C—*Transformer fingerprints.*

Gas, ppm (Vol./Vol.)	Example 1		Example 2	
	I	3	I	3
H_2	0	1	0	0
CH_4	92	69	15	18
CO	370	400	33	57
C_2H_6	2300	2300	560	520
CO_2	6000	6800	1800	2200
C_2H_4	180	180	9	6
C_2H_2	0	0	0	0

NOTE: I = initial.

However, it should be noted that in each of the two data sets, the first row of values represents the initial results obtained on the "new" transformer while the second row is the corresponding set of gas values three years later. Consequently, one must reconsider since the gas values three years later are nearly the same. It appears that: (1) there apparently is no problem because the units are still in service; (2) the distribution of gases represents an equilibrium condition since the values over three years are nearly constant; and (3) the predominance of hydrogen and carbon monoxide is a unique condition. As a matter of fact, these transformers are from the same manufacturer and the gas data appear to be the normal attributes of a particular class of distribution transformers.

In Table 8B, another two sets of dissolved gas results are presented in which the predominant gases are ethane and ethylene. Again, as in the previous table, the first row shows the initial values and the second row shows the respective gas values two years later.

Considering the predominant gases in this case, ethylene and ethane, along with a detectable amount of acetylene appearing also in the first group, one might again entertain the thought of a possible problem, at least in the first group.

Since the predominant (key) gases are relatively constant over a two-year period, it appears as in the previous situation that the dissolved gas compositions represent an equilibrium condition which constitutes a "norm" for these transformers. Actually, the equipment represented by this gas data also is from a single source, but different from those depicted in Table 8A. The first group with the greater ethylene content applies to generator step-up transformers and the second group to substation units.

Table 8C provides another example where somewhat higher gas levels are apparent, but the conclusions are the same.

Transformer "fingerprints" are a means of establishing normal trends for transformers based on examination of many transformers of the same manufacturer and design.

Transformer dissolved gas "fingerprints" may not always be as distinct or revealing, for consideration must be given to the modifying influences of service conditions, for example, loads and temperature, in addition to rating and venting. Fingerprinting obviously is enhanced when service conditions are more uniform.

CO/CO_2

Cellulose insulation yields mostly CO and CO_2 when subjected to conditions which result in its deterioration. Norms, generation rates, and the ratio of CO to CO_2, or CO_2 to CO, have been used as an indicator of abnormal cellulose degradation. Normal values of CO and CO_2 are given in Table 3. McNutt et al. have developed a correlation between CO gassing rate and temperature and surface area involved [11]. Rogers [24] determined the average rate of CO formation in a population of generation transformers as:

$$\text{ppm}\ (V/V) = 374\ \text{Log}_{10}\ 4\,Y\ (\text{where}\ Y = \text{age in years})$$

The use of CO/CO_2 ratios suffers from the presence of "accidental" CO_2 from air and from variations in O_2 concentration present in the transformer incipient fault area (high O_2 favors formation of CO_2). Also, oil, if sufficiently heated in the presence of excess oxygen, will also generate some CO and CO_2. Experiments by Vitols and Fernandes [30] show that ratios of CO to CO_2 increase with increasing temperature and that, based on population studies, "normal ratios" range from 0.07 to 0.30. Rogers [24] found that the average CO_2/CO ratio on 1000 generator transformers was 7/1 ($CO/CO_2 = 0.14$) with a standard deviation of 4. Extreme ratios of CO/CO_2, approaching unity or greater than 1/15, should be considered as unusual behavior, with the unity ratio being of more serious concern as high temperatures/electrical faults may be involved. For example, in a case given by Vieira [27], where severe arcing is apparent (high C_2H_2—5390 ppm) and cellulose appears to be involved (high CO—1130 ppm), the ratio of CO to CO_2 was 1130/285 or 4/1.

Water

Much effort goes into reducing the water content and maintaining low concentrations of water in the transformer insulation due to its obvious deleterious effects. However, in some cases where leaks are present or large operating temperature swings occur causing supersaturation and formation of free water, water is converted in the presence of core steel and heat to H_2 and O_2 [31]. The electrochemical breakdown of free water can also occur. Wet cellulose under corona discharge condition will yield more H_2 than dry cellulose [32]. It is important to monitor water concentration to distinguish the cause of H_2 formation which may be due to the presence of free water or corona activity (dissolved water only).

Other Materials

Power transformers are mostly composed of insulating oil, cellulose paper and pressboard, conductor, and core steel. Some other constituents used in lesser percentages, such as silicone compounds, organic polymers, etc. may also be subject to degradation under severe thermal and electrical discharge conditions. In most cases, these other materials under arcing or pyrolysis conditions generate predominantly H_2 and CO, although hydrocarbons, CO_2, and water may also be generated in some cases or under certain conditions [33,34].

Other Fluids

In special applications alternative fluids such as silicone fluids are used instead of mineral oils. As these fluids are used, diagnostic schemes need to be developed from population studies, laboratory simulations, and examination of failed transformers, as has been done for mineral oils. Laboratory simulations on silicone fluids have revealed the diagnostic scheme given in Table 9 [35].

As would be expected, silicone transformer fluid (polydimethyl siloxane) yields lesser quantities of C_2s than mineral oils under the same conditions, although the total combustible gases generated are of the same order.

It should be noted that air is more soluble in silicone fluids than mineral oils and, with high oxygen present with overheating conditions not involving the conductor, the predominant gas generated is CO with lesser quantities of CH_4 and C_2H_2 found. It should still be possible in most cases to distinguish between overheating in silicone versus overheating in paper because the formation of CH_4 and C_2H_4 occurs in significant quantities only in the case of overheating in silicone. Also, in the case of a sustained arcing, silicone fluids generate a small percentage of C_2H_2 relative to the other combustible gases and, therefore, the presence of any acetylene in a silicone transformer should be considered as very serious. The percent of C_2H_2 in the total combustible gas in mineral oil and silicone fluid under arcing conditions is about 50% and 3%, respectively.

Little information is known about field experience with silicone transformers as they have not been in service in large numbers in the United States until recent years. As service experience is gained, norms can be developed for silicone transformers as has been done for mineral oil units.

Conclusions

Dissolved gas-in-oil analysis has been a useful tool for detecting problems in power transformers which are not detectable by other means or in confirming problems indicated by other tests. A comprehensive scheme is required for diagnostic purposes due to the many variables which determine or affect the types and quantities of gases generated. (Prudent judgement is also an integral part of the final analysis, particularly when incipient faults are indicated.) Historical and nameplate data and the type of transformer oil preservation system supply important information used in diagnosing transformer health. Characteristic combustible gases generated by incipient faults are determined and used in interpretive schemes consisting of norms (acceptance levels), total combustible gas levels, key gases, ratios, gassing rates, trends, and "fingerprints."

Norms, total combustible gas, key gases, and ratios (for ancillary information) are used for the analysis of first-time samples to establish the future sample frequency. Subsequent analysis incorporates total gas, key gases, ratios, trends, gassing rates, and "fingerprints" to diagnose transformer health.

The future diagnostic schemes for dissolved gas-in-oil analysis should include more definition of norms for individual transformer manufacturers' designs ("fingerprints") based on large

TABLE 9—*Diagnostic gases dissolved in silicone fluid.*

Condition	Key Gases (Listed in Order of Highest to Lowest Concentration)
Overheating	CH_4(CH_4, CO, C_2H_4, and H_2 in presence of copper) CO, CH_4, C_2H_4 in presence of high O_2
Corona	H_2, CH_4
Arcing	H_2, CH_4, CO, C_2H_2

data bases. A comprehensive system such as the use of artificial intelligence (AI), which has already started [36], may be used to diagnose simple cases (90%). Further refinement of the AI system should enable it to diagnose the more complex problems as has been done, for example, in the medical field.

For computerized methods, a comprehensive system such as AI is needed to handle the complexity of information necessary to diagnose incipient fault conditions and to supply the information used in drawing a conclusion when requested.

References

[1] ASTM D 3612, Method for Analysis of Gases Dissolved in Electrical Insulating Oil by Gas Chromatography, *Annual Book of ASTM Standards, Vol. 10.03, Electrical Insulating Liquids and Gases; Electrical Protective Equipment,* ASTM, 1916 Race St., Philadelphia, PA, 1986.

[2] Baker, A. E., "Solubility of Gases in Transformer Oil," Minutes of the Forty-Sixth Annual International Conference of Doble Clients, 1979, Sec. 10-701, Doble Engineering Co., Watertown, MA.

[3] Oommen, T. V., "Adjustments to Gas-in-Oil Analysis Data Due to Gas Distribution Possibilities in Power Transformers," *IEEE Transactions,* Vol. PAS 101, No. 6, June 1982, pp. 1716–1722.

[4] ASTM D 2779, Standard Method for Estimation of Solubility of Gases in Petroleum Liquids, *Annual Book of ASTM Standards, Vol. 5.02, Petroleum Products and Lubricants (II),* ASTM, 1916 Race St., Philadelphia, PA, 1986.

[5] Duval, M. "Fault Gases Formed in Oil-Filled Breathing EHV Power Transformers—The Interpretation of Gas Analysis Data," *IEEE Transactions,* PAS C.74 476-8, 1974.

[6] McNutt, W. J. et al., "Mathematical Modelling of Bubble Evolution in Transformers," *IEEE Transactions,* Vol. PAS 104, No. 2, Feb. 1985, pp. 477–87.

[7] Chadwick, A. T. et al., "Oil Preservation Systems; Factors Affecting Ionization in Large Transformers," *AIEE Transactions,* April 1960, pp. 92–99.

[8] Blanchard, C. S., "Pownal Substation No. 1 Bank Oil Bubbling Problem," ECNE Fall Meeting, 1985, Electric Council of New England, Bedford, MA.

[9] Kaufmann, G. H., "Gas Bubbles in Distribution Transformers," *IEEE Transactions,* Vol. PAS 96, No. 5, Sept./Oct. 1977, pp. 1596–1601.

[10] Heinrichs, F. W., "Bubble Formation in Power Transformer Windings at Overload Temperatures," *IEEE Transactions,* Vol. PAS 98, No. 5, Sept./Oct. 1979, pp. 1576–81.

[11] McNutt, W. J. et al., "Short-Time Failure Mode Considerations Associated with Power Transformer Overloading," *IEEE Transactions,* Vol. PAS 99, No. 3, May/June 1980, pp. 1186–97.

[12] Baker, A. E. et al., "An Update on Fault Gas Analysis (A Review)," Minutes of the Fifty-Third Annual International Conference of Doble Clients, 1986, Sec. 10-301, Doble Engineering, Co., Watertown, MA.

[13] Doble Client Committee on Insulating Fluids, "Combustible Gas Content in the Oil of New Transformers Prior to Initial Energization," Minutes of the Fiftieth Annual International Conference of Doble Clients, 1983, Sec. 10-701, Doble Engineering Co., Watertown, MA.

[14] Oommen, T. V. et al., "Experience with Gas-in-Oil Analysis Made During Factory Tests on Large Power Transformers," *IEEE Transactions,* Vol. PAS 101, No. 5, May 1982, pp. 1048–52.

[15] Pugh, D. R., "Combustible Gas Analysis," Minutes of the Fortieth Annual International Conference of Doble Clients, 1973, Sec. 10-401, Doble Engineering Co., Watertown, MA.

[16] Pugh, D. R., "Advances in Fault Diagnosis by Combustible Gas Analysis," Minutes of the Forty-First Annual International Conference of Doble Clients, 1974, Sec. 10-1201, Doble Engineering Co., Watertown, MA.

[17] Hays, T. W., "Investigation to Determine the Location of a Low-Energy, Audible Electrical Arcing in a Power Transformer," Minutes of the Fifty-Third Annual International Conference of Doble Clients, 1986, Sec. 6-301, Doble Engineering Co., Watertown, MA.

[18] Oms, L. W., "Transformer Problems Detected or Confirmed by Use of Combustible Gas Analysis (A Progress Report)," Minutes of the Forty-Eighth Annual International Conference of Doble Clients, 1981, Sec. 10-101, Doble Engineering Co., Watertown, MA.

[19] Marquez, A., "Recurring Core Ground Found by Combustible-Gas Analysis," Minutes of the Forty-Fourth Annual International Conference of Doble Clients, 1977, Sec. 6-301, Doble Engineering Co., Watertown, MA.

[20] Halstead, W. D., "A Thermodynamic Assessment of the Formation of Gaseous Hydrocarbons in Faulty Transformers," *Journal of the Institute of Petroleum,* Vol. 59, Sept. 1959, pp. 239–241.

[21] Manger, H. C., "Combustible Gas Ratios and Problems Detected," Minutes of the Forty-Fifth Annual International Conference of Doble Clients, 1978, Sec. 6-1101, Doble Engineering Co., Watertown, MA.

[22] Rickley, A. L. et al., "Analysis Techniques for Fault-Gas Analyses (A Progress Report)," Minutes of the Forty-Fifth Annual International Conference of Doble Clients, 1978, Sec. 10-401, Doble Engineering Co., Watertown, MA.

[23] Dornenburg, E. and Strittmatter, W., "Monitoring Oil-Cooled Transformers by Gas Analysis," *Brown Boveri Review*, Vol. 5, No. 74, pp. 238–247.

[24] Rogers, R. R., "U.K. Experience in the Interpretation of Incipient Faults in Power Transformers by Dissolved Gas-in-Oil Chromatographic Analysis (A Progress Report)," Minutes of the Forty-Second Annual International Conference of Doble Clients, 1975, Sec. 10-201, Doble Engineering Co., Watertown, MA.

[25] Rogers, R. R., "U.K. Experience in the Interpretation of Incipient Faults in Power Transformers by Dissolved Gas-in-Oil Chromatographic Analysis (A Progress Report)," Minutes of the Forty-Fourth Annual International Conference of Doble Clients, 1977, Sec. 10-50, Doble Engineering Co., Watertown, MA.

[26] Rogers, R. R., "Concepts Used in the Development of the IEEE and IEC Codes for the Interpretation of Incipient Faults in Power Transformers by Dissolved Gas-in-Oil Analysis," presented at IEEE 1978 Winter Power Meeting.

[27] Sobral Vieira, C. L. C., "Correlation Between Results of Dissolved Gas Analysis and Actual Transformer Inspection," Minutes of the Fifty-Third Annual International Conference of Doble Clients, 1986, Sec. 6-501, Doble Engineering Co., Watertown, MA.

[28] Lyke, A. J. and Vitols, A. P., "Automated Monitoring of Dissolved Gas-in-Oil for Large Power Transformers," Minutes of the Forty-Fourth Annual International Conference of Doble Clients, 1977, Sec. 10-601, Doble Engineering Co., Watertown, MA.

[29] MacDonald, J. D. and Vitols, A. P., "Gas-in-Oil Analysis as a Diagnostic Tool for Monitoring Power Transformer Insulation Integrity," Minutes of the Forty-Seventh Annual International Conference of Doble Clients, 1980, Sec. 6-901, Doble Engineering Co., Watertown, MA.

[30] Vitols, A. P. and Fernandes, R. A., "Incipient Fault Detection for EHV Transformers," IEEE Summer Power Meeting, Vancouver, BC, July 15–20, 1979, IEEE, New York.

[31] Sheppard, H. R., "The Mechanism of Gas Generation in Oil-Filled Transformers," Minutes of the Thirtieth Annual International Conference of Doble Clients, 1963, Sec. 6-601, Doble Engineering Co., Watertown, MA.

[32] Baker, A. E., "Gas Composition in Corona Discharge," Minutes of the Forty-Ninth Annual International Conference of Doble Clients, 1982, Sec. 10-701, Doble Engineering Co., Watertown, MA.

[33] Pugh, D. R. and Belanger, G., "Pyrolysis of Insulating Materials," Minutes of the Forty-Fourth Annual International Conference of Doble Clients, 1977, Sec. 10-901, Doble Engineering Co., Watertown, MA.

[34] Hettwer, P. F., "Arc-Interruption and Gas-Evolution Characteristics of Common Polymeric Materials," *IEEE Transactions*, Vol. PAS 101, No. 6, June 1982, pp. 1689–1696.

[35] Griffin, P. J., "Analysis for Combustible Gases in Transformer Silicone Fluids," Minutes of the Fifty-Second Annual International Conference of Doble Clients, 1985, Sec. 10-701, Doble Engineering Co., Watertown, MA.

[36] Lowe, R. I., "Artificial Intelligence Techniques Applied to Transformer Oil Dissolved Gas Analysis," Minutes of the Fifty-Second Annual International Conference of Doble Clients, 1985, Sec. 10-601, Doble Engineering Co., Watertown, MA.

DISCUSSION

C. Vieira[1] *(written discussion)*—What is the policy in the United States in using dissolved gas-in-oil tests after factory heat runs? Is there any consideration by ASTM to modify Method D 3612 based on this application?

It is important to remember that all the diagnostic criteria for dissolved gas-in-oil analysis were developed to monitor the condition of in-service transformers. With factory tests the con-

[1]CEPEL, P.O. Box 2754, Rio de Janeiro, Brazil.

[2]Musil, J. and Foschum, H., "Application of Dissolved Gas Analysis (DGA) During Factory Testing of Power Transformers," Annual International Conference of Doble Clients, Sec. 6-801, 1980, Doble Engineering Co., Watertown, MA.

centration of each gas is very small, and sometimes a variation of 2 ppm corresponds to more than a 100% increase.

P. Griffin (author's closure)—In the United States there are no standard limits for dissolved gases in oil generated during heat runs performed in the factory. However, manufacturers have developed guidelines based on their experience as shown in Table 4 and in information in Refs *13*, *14*, and in Footnote 2. Often these limits are divided into two categories. The first set of limits indicates that a problem may exist and that more testing should be performed. A second set of higher limits indicates that a definite problem exists and that a thorough investigation should be performed. Some utilities in the United States specify that dissolved gas-in-oil tests must be performed after the heat run. Limits may or may not be specified.

ASTM Method D 3612 limits of detection for gases dissolved in oil are:

Gas	Minimum Detection Limit, ppm
Hydrogen	5
Hydrocarbons	1
Carbon oxides	25
Atmospheric gases	50

Committee D-27, which is responsible for this test method, is currently examining the method to evaluate what factors can be controlled to improve accuracy and precision. Although the performance of this work was not instigated by concern for factory heat run tests, it certainly would be most important in this case, where small numbers and changes are being evaluated. The limits of detection of the test method are not presently being examined.

T. J. Haupert[1] *and Fredi Jakob*[1]

A Review of the Operating Principles and Practice of Dissolved Gas Analysis

REFERENCE: Haupert, T. J. and Jakob, F., **"A Review of the Operating Principles and Practice of Dissolved Gas Analysis,"** *Electrical Insulating Oils, STP 998,* American Society for Testing and Materials, Philadelphia, 1988, pp. 108–115.

ABSTRACT: Dissolved gas analysis has gained worldwide acceptance as a method for the detection of incipient faults. When properly applied, the method provides a highly reliable and sensitive method for the identification of fault type and for the assessment of the rate of fault development.

This paper investigates the origins of the individual fault gases, factors that are responsible for the extreme sensitivity of the method, and the factors that influence the observed gas concentrations. The paper also examines the scope and limitations of the diagnostic techniques that have been developed to interpret the data that the method provides. These techniques include such concepts as key gases, ratios of gas concentrations, nomographs, and rates of gas generation.

KEY WORDS: dissolved gases, fault detection, incipient faults, gas chromatography, key gases, gas concentration ratios, nomograph, gas generation rates

Fifteen years ago when Analytical Associates began its dissolved gas analysis (DGA) program, it was not generally recognized that DGA could be a useful diagnostic and preventive maintenance tool for the detection of faults in oil-filled electrical equipment. DGA has, in the ensuing years, been widely accepted by the power industry, and it is currently recognized as a significant diagnostic tool for the detection of faults in oil-filled transformers. The wide acceptance of DGA is based on its sensitivity, which is orders of magnitude greater than other gas detection methods, and its reliability. As a result, faults can be detected and identified at the earliest possible time, and the method provides an excellent means of monitoring incipient fault development.

Fault gases are produced by degradation of the transformer oil and other insulating materials such as cellulose. In the presence of an active fault, the rate of oil and cellulose degradation is significantly increased, and the types of degradation products formed will vary with the nature and severity of the fault. The composition of the gas mixture resulting from the decomposition is energy dependent, and, since fault processes differ greatly in the energy they dissipate, different mixtures of degradation products are produced for each type of fault process. By observing the composition of the gases produced by the degradation of the insulating media, it is possible to distinguish three basic fault processes which differ greatly in their energy characteristics. These processes are arcing, corona or partial discharge, and pyrolysis or thermal decomposition. Pyrolysis occurs whenever there is sufficient thermal energy to break chemical bonds. Pyrolysis can occur in the absence of oxygen and produces different products than a combustion process. Thermal decomposition is a similar process but usually implies lower temperatures and lower reaction rates. The low-molecular-weight gases that are formed vary widely in terms of their equilibrium solubilities in oil, from a low of 7% for hydrogen to a high of 400% on a volume/volume basis for acetylene. But one must realize that even a solubility of 7% is very high in terms of the concentrations of those gases that actually occur in a fault. Therefore, the oil is rarely saturated with the fault gases. Coupled with the fact that the volume of oil is significantly

[1]Analytical Associates, Inc., Sacramento, CA 95826.

greater than the volume of the gas blanket, if one is present, it is apparent that most of the fault gases remain dissolved in the oil. Thus, it is far better to analyze an oil sample for the presence of a fault gas rather than to look in the gas above the oil. The sensitivity of DGA is also better than that of other methods of gas detection such as portable or fixed space gas monitors and pressure-sensing devices. Using DGA, fault gases can be routinely detected at the part-per-million (ppm) level. Thus, DGA is often successful in identifying transformer problems before there is an indication of a problem with other test procedures.

Buchholtz and others were aware of the production of fault gases in transformers and tried as early as 1928 to correlate the gases produced with a particular type of fault [1]. Historically, two different approaches have been used to try to correlate fault gas composition and fault type. One approach is empirical and is based on statistical evaluation of fault gas data from units that have exhibited documented problems. The Central Electricity Generating Board of Great Britain (CEGB) has been a leader in this area [2]. They started extensive work on fault-gas correlations as early as 1968. The nonempirical approaches have treated the production of fault gases from both an experimental and thermodynamic approach. Experimental data have shown that an increase in oil temperature led to an increase in the ratio of unsaturated to saturated hydrocarbons, specifically, the ratios of acetylene to ethylene and ethylene to ethane. Halstead [3], using a thermodynamic analysis, predicted that low-molecular-weight fault gases would be produced with increasing temperature in this sequence: methane, ethane, ethylene, and finally acetylene.

The empirical and theoretical studies have led to similar conclusions. The consensus is that certain key gases can be correlated with fault type and that the rate of gas production can be correlated with fault severity. It is generally accepted that acetylene can only be produced at the very high temperatures that occur in the presence of an arc. All other low molecular weight hydrocarbons, when produced concurrently with hydrogen, indicate pyrolysis of the oil. Corona or partial discharge is a relatively low temperature process and produces mainly hydrogen.

Dornenburg and Strittmatter [4] are generally credited with the development of the gas ratio concept. They noted that fault gas ratios clustered according to fault type when plotted on a log-log graph. Their plots are reproduced in Fig. 1. These investigators also reported gas levels that were observed in transformers that have been operating free of faults for several years. These normal gas values are tabulated in Table 1. Subsequent experience [5] has shown that levels slightly higher than those recommended by Dornenburg and Strittmatter are applicable to American transformers. The difference in normal levels between European and American transformers can be ascribed to differences in design and operating parameters. The higher levels recommended by the Bureau of Reclamation are tabulated in Table 2. R. R. Rogers, a CEGB engineer, extended the ratio concept and published interpretive guides that facilitated the interpretation of gas ratios [6].

The pioneering work of Dornenburg, Strittmatter, and Rogers led to the development of a nomograph for the interpretation of DGA data. J. O. Church of the U.S. Department of the Interior provided the leadership in this work, tested and evaluated the nomograph and coauthored a paper with us which describes the development of the nomograph [5]. Our objective in developing the nomograph was to combine the concepts of fault gas threshold levels and the gas-ratio method in a convenient manner to facilitate interpretation of DGA data. Fault gas concentrations are plotted in ppm on several vertical logarithmic scales as shown in Fig. 2. Lines are then drawn to connect the concentration values on adjacent scales. Points higher on the scales are given greater significance than lower points. The unshaded tick marks on each scale represent the threshold criteria of Dornenburg and Strittmatter. Our experience has shown that the concentration limits can be raised as noted by the shaded symbols. The slope of the line connecting the two adjacent concentrations is thus directly related to the fault gas concentration ratio. Figure 3 illustrates the method as applied to one of the gas pairs, methane and hydrogen. For this gas pair, connection of any set of points representing equal concentrations of hydrogen and methane generates a horizontal line. Thus, a horizontal line corresponds to a methane to

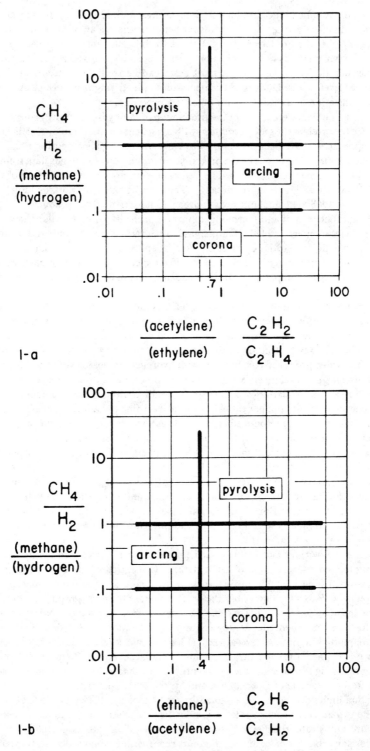

FIG. 1—*Dornenburg and Strittmatter's plots of fault gas ratio pairs.*

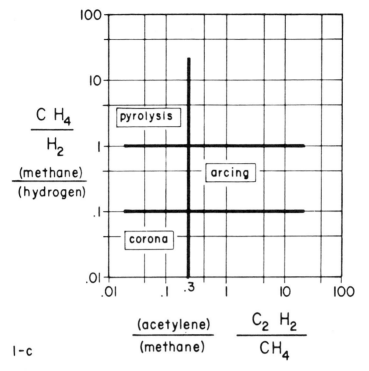

I-c

FIG. 1—*Continued.*

TABLE 1—*Dornenburg and Strittmatter suggest that transformers which have been in service for a few years may be regarded as operating without problems if gas concentrations do not exceed the following parts-per-million levels.*

Hydrogen	200
Methane	50
Ethane	35
Ethylene	80
Acetylene	5
Carbon monoxide	500
Carbon dioxide	6000

TABLE 2—*Threshold levels utilized by the Bureau of Reclamation.*

Hydrogen	500
Methane	125
Ethane	75
Ethylene	175
Acetylene	15
Carbon monoxide	750
Carbon dioxide	11 000

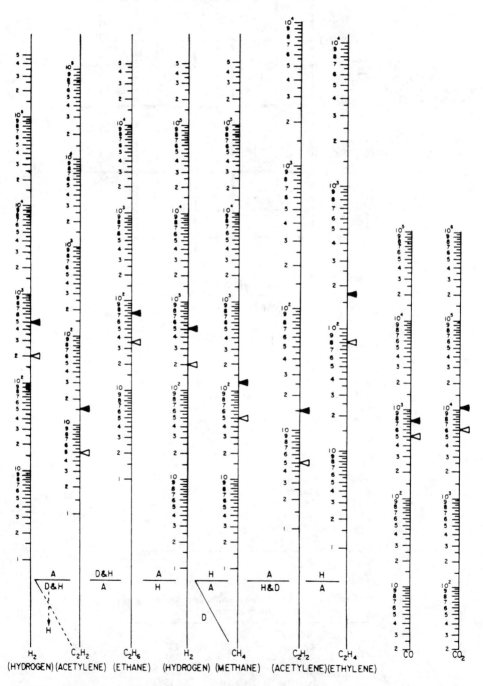

FIG. 2—*The Church nomograph. Open triangles indicate threshold levels for fault gases according to Dornenburg and Strittmatter. Filled triangles indicate threshold levels currently utilized by the Bureau of Reclamation.*

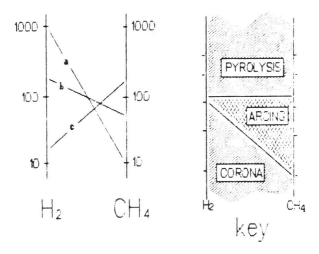

FIG. 3—*The interpretation of hydrogen to methane ratios. An illustration of how the Church nomograph was developed.*

hydrogen ratio of one. A ratio greater than one gives a line with a positive slope and indicates pyrolysis as the fault mode. Negative slopes in the range of 0 to -1, corresponding to a ratio of hydrogen to methane in the range of 1 to 10, indicate arcing. Negative slopes greater than -1, corresponding to a hydrogen to methane ratio greater than 10, indicate corona or partial discharge. Figure 3 illustrates these conditions. For example, if the oil contains 1000 ppm of hydrogen and 10 ppm of methane, the line has a negative slope (Line a), and the failure mode indicated is corona. If the DGA revealed a hydrogen concentration of 200 ppm and 70 ppm of methane (Line b) the connecting line would have a negative slope and would indicate arcing. If the methane concentration exceeds the hydrogen concentration, a line with a positive slope results. Such would be the case if the methane concentration was 200 ppm while the hydrogen concentration was 20 ppm. This line (c) indicates pyrolysis as the failure mode. It is important to note that in each of these examples at least one of the pair of gases was above the threshold level.

Carbon monoxide and carbon dioxide are indicative of cellulose decomposition in a transformer. Dornenburg and Strittmatter suggest threshold levels of 500 ppm for carbon monoxide and 6000 ppm for carbon dioxide. The Bureau of Reclamation, based on their experience, has raised these levels to 750 ppm of carbon monoxide and 10 000 ppm for carbon dioxide. Concentrations exceeding these values indicate that cellulose material is involved in the fault process. A negative slope for the line connecting these two concentrations indicates a carbon monoxide to carbon dioxide ratio greater than 0.1. This ratio is indicative of an accelerated solid insulation decomposition rate which can be ascribed to higher temperatures associated with a fault.

Application of the nomograph can be illustrated by what we consider a classic case. The results shown in Fig. 4 are from a mobile transformer that was being tested for the effects of operating without the cooling system functioning. During shutdown an internal ground fault occurred. DGA samples were fortunately taken before and immediately after the occurrence of the fault. The before and after results are tabulated in Table 3. An immediate and significant increase of all the fault gases is obvious. The results again illustrate the sensitivity and ability of DGA to detect a problem at the earliest possible time, in this case minutes after the fault occurred. It is also of interest to note how consistently the various gas ratios indicate that an arc occurred in this mobile transformer. The advantages of using the nomograph are summarized in Table 4.

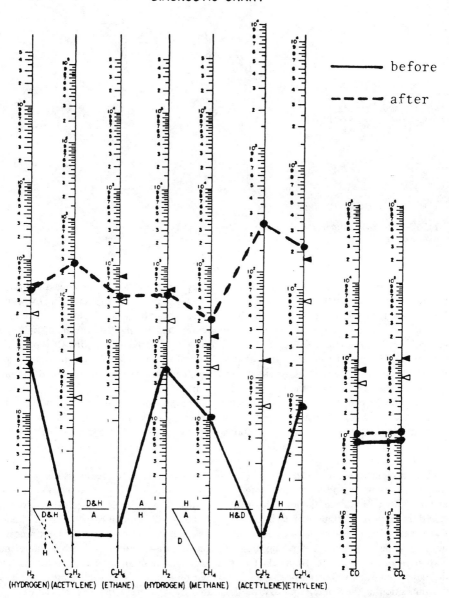

DISSOLVED GAS-IN-OIL ANALYSIS
DIAGNOSTIC CHART

FIG. 4—*A Church nomograph plot of the data in Table 3.*

Although the theoretical basis of DGA is well established, several problems exist in the current practice of the technique. These problems are related not only to the interpretation of DGA results but also to the generation of the results themselves. The determination of dissolved gases is usually a two-step laboratory process. In the first step, the dissolved gases are separated from the transformer oil sample by extraction into an evacuated gas collection apparatus. The efficiency of this extraction is dependent upon the initial vacuum achieved in the system, the volume of the evacuated system into which the gases diffuse, and the length of time allowed for the

TABLE 3—*Before and after data for mobile transformer (see Fig. 4) in parts per million (ppm).*

Gas	Before	After
Hydrogen	45	400
Acetylene	0	270
Ethane	0	43
Methane	10	200
Ethylene	8	230

TABLE 4—*The use of the nomograph has been found to offer distinct advantages in analyzing data obtained from DGA. These advantages can be easily summarized.*

1. Data can be quickly and easily interpreted.
2. High gas levels are immediately apparent.
3. Relative changes in concentrations are easily seen.
4. The rate of change in gas evolution can be observed when subsequent tests are plotted on the same nomograph.

extraction. The dissolved gases are then compressed back to atmospheric pressure. Some laboratories compress the gas in the presence of the oil, which necessitates a correction to be applied for the reabsorption of the extracted gas into the oil. The volume of the collected gas is measured and corrected to standard temperature and pressure (STP). An aliquot of the extracted gas is then analyzed qualitatively and quantitatively by gas chromatography (GC). Here again there is a wide range of procedures employed. Some laboratories use a single chromatographic column, others use two with two detectors. Some employ a single sample injection, others inject two aliquots of the sample. Thermal-conductivity detectors are used by some laboratories; others use flame ionization detectors. Samples of standard gas mixtures are available so that the analyses of the extracted gases can be correctly made despite differences in the analytical systems. However, standards of oil containing dissolved gases are not presently available. As a result, differences in the extraction efficiencies of the gases among laboratories will affect the calculated results of the concentration of the individual gases dissolved in the oil. If greater interlaboratory consistency of results is to be achieved, two things are necessary. First, laboratories must agree to use the ASTM Method for Analysis of Gases Dissolved in Electrical Insulating Oil by Gas Chromatography (D 3612) or a revised version of this standard method, and, secondly, chemical standards of oil containing dissolved gases must be made available. Despite differences that exist among laboratories in the extraction efficiencies that affect the reported concentrations of dissolved gases, little or no problem exists if all analyses are performed using the same analytical system. Also, the effect of differences in the extraction efficiencies will be minimized if ratios of gas concentrations are used for diagnostic purposes, rather than the individual gas concentrations.

References

[1] Buchholtz, M., *Electrische Technichse Zeitschrift,* Vol. 49, No. 34, 1928, pp. 239–241.
[2] Rogers, R. R., *IEEE Transaction Electrical Insulations,* Vol. EI-13, No. 5, October 1978.
[3] Halstead, W. D., *Journal of the Institute of Petroleum,* Vol. 59, No. 569, September 1973, pp. 239–241.
[4] Dornenburg, E. and Strittmatter, W., *Brown Boveri Review,* Vol. 61, No. 5, 1974, pp. 238–247.
[5] Church, J. O., Haupert, T. J., and Jakob, F., *Electrical World,* Vol. 201, No. 10, October 1987, pp. 40–44.
[6] Rogers, R. R., Doble Client Conference 1977, Paper 44 AIC 77, Doble Engineering, Boston, MA.

Section V—Electrostatic Buildup in Transformer Oil

J. Franklin Roach[1] and James B. Templeton[2]

An Engineering Model for Streaming Electrification in Power Transformers

REFERENCE: Roach, J. F. and Templeton, J. B., **"An Engineering Model for Streaming Electrification in Power Transformers,"** *Electrical Insulating Oils, STP 998,* H. G. Erdman, Ed., American Society for Testing and Materials, Philadelphia, 1988, pp. 119–135.

ABSTRACT: An engineering approach to assessing static electrification phenomena in power transformers is presented. A theoretical model is developed to estimate streaming currents and space charge accumulation around the hydraulic oil flow loop in a three-phase transformer. The streaming current model is applied to a 345-kV, 240-MVA three-phase auto transformer with shell form winding design. The influence of space charge on the electric field and potential distributions in the upper plenum is modelled by a one-dimensional theory. Significant electric stress enhancement at coil edges on alternate half cycles of applied a-c stress is predicted for the 240-MVA transformer. Potential failure modes in power transformers due to static electrification are discussed.

KEY WORDS: transformers, power, electrification, streaming, current, shell form, core form, charge, static, failure, oil, paper

In the last six years considerable attention has been directed at investigating streaming electrification in power transformers. Power transformers are constructed with either shell form or core form configurations and employ an insulation system comprised of cellulosic solid materials with mineral oil as an insulating fluid and heat transfer medium.

As the MVA throughput of transformers increases, it becomes necessary to provide cooling of the core and coil assembly by means of forced oil and forced air (FOA) heat transfer. The FOA cooling is accomplished by flowing oil through the core and coil assembly, out of the transformer tank, through externally mounted oil to air heat exchangers, and back into the transformer tank in a closed loop. In a typical elevation cross section of a shell form power transformer, for example, the oil is injected into a lower plenum via pumps and is directed vertically through the core and coil assembly. The oil continues to flow vertically through a lead support region above the core and coils referred to as the bridge or superstructure and then enters the upper plenum, where the oil is drawn out of the tank and into the external heat exchangers completing the loop at the pump.

The existence of an insulating fluid flowing past an insulating solid results in charge separation at the interface of the two materials. Physically, charges separate at the interface of the oil and paper in any transformer regardless of construction type, that is, shell form or core form. The importance of studying the streaming electrification phenomena is to enable one to predict when the magnitude of streaming electrification becomes problematical. It is known that the streaming electrification phenomena has been a major contributor to a number of failures of large transformers in service. Therefore, it is necessary to understand the theory of streaming electrification so that design criteria may be developed to avoid the problematic effects of streaming electrification. The purpose of this paper is to explain some of the theory embodied in

[1]Senior research scientist, Research and Development Center, Westinghouse Electric Corp., Pittsburgh, PA 15235.

[2]Manager, Shell Form Design, Power Equipment Division, Westinghouse Electric Corp., Pittsburgh, PA 15235.

flow electrification in power transformers and to show a mathematical model which may be used to predict the magnitude of the streaming currents and the impact they have on voltage stresses within the transformer.

Flow Electrification

When a low conductivity liquid, such as transformer oil, flows past a solid surface, the liquid acquires a space charge due to frictional charge separation at the liquid-solid interface. In general, ionic species of one sign are preferentially absorbed by the solid, leaving a net charge of opposite sign in the liquid. The space charge developed in the liquid is transported by the flow, resulting in a streaming current or charging current being carried by the liquid. If the walls of the flow system are insulated or floating, the flow electrification process also leads to an accumulation of charge and the generation of high electrostatic surface potential at liquid-solid interfaces. The phenomenon of charge generation by low-conductivity liquids flowing in pipes and ducts is well known in the petroleum industry. A number of explosions and fires have been attributed to discharges initiated in fuel tanks and handling equipment due to accumulation of charge generated by flow electrification.

Electrostatic charge generation is also known to occur in forced oil cooled power transformers. As the cooling oil flows through the insulation structure, charge separation takes place at oil/paper interfaces. Charge accumulation results both in the oil and on insulating surfaces, producing d-c fields which are superimposed on the a-c energization field. If static charge fields are high enough, spontaneous partial discharges will occur in the transformer, even when unenergized. It is also possible that impulse and switching surges of the right polarity in combination with the static charge potentials may produce localized stresses of sufficient magnitude to initiate catastrophic failure of the insulation. Therefore, it is important to investigate and quantify the phenomena of static charge generation in transformers.

The subject of flow electrification has been most extensively studied in the petroleum industry. The much quoted work of Klinkenberg and Van der Minne [1] published in 1958 gives an early account of electrostatics related to the transport of petroleum liquids in pipes and the hazards of charge accumulation in storage tanks. In 1965 a study of charge separation in liquid hydrocarbons was conducted by Shafer, Baker, and Benson [2] at the National Bureau of Standards. More recent work on static charge in petroleum liquid has been reported by Itoh [3].

The problem of static electrification in both shell form and core form power transformers was recognized in the early 1970s. Notable papers on flow electrification in transformers have been published by a number of Japanese authors, including Kan [4], Shimizu [5], Higaki [6], Okubo [7], Tanaka [8], and Tamura [9]. A CIGRE paper by Takagi [10] on reliability improvement of 500-kV power transformers also discusses the static charge problem. The static electrification properties of oil have been studied by Oommen [11,12]. In an important recent paper by Crofts [13], the static electrification phenomena in power transformers is reviewed in detail. Crofts discusses the failures of several power transformers which were attributed to static electrification phenomena.

Theoretical treatments of flow electrification for low conductivity fluids have been published by Klinkenberg and Van der Minne [1], Gavis and Koszman [14-16], Schön [17], Abedian and Sonin [18], and Tanaka, Yamada, and Yasojima [19]. The paper by Tanaka et al. addresses the electrification theory specifically for transformer insulation and gives empirical streaming current formulations for both laminar and turbulent flow regimes.

The primary factors which influence the flow electrification process in a forced oil cooled transformer are: (1) the oil flow rate; (2) the oil bulk oil properties of temperature, electrical conductivity, and viscosity; (3) surface conditions and structure of the solid insulating materials of the oil ducts; and (4) moisture content of oil and paper. Other important factors are the nature of the ionic content of the oil, which is influenced by the refining process; the time history of oil as it circulates in the transformer; and contamination particles. The charging tendency of a transformer oil is therefore a strong function of the ionic adsorption chemistry which occurs at oil-paper interfaces. The net rate of adsorption of charges by paper surfaces will be dependent upon oil deterioration products such as copper and aluminum ions, polar molecules, etc., and upon physical and chemical changes of the insulation materials themselves. The charge separation process is also dependent upon the application of voltage to the transformer windings. In general it is found that a-c energization leads to enhanced charge separation.

In this paper a flow electrification model for a three-phase, shell form transformer is developed. The model is applied to a 345-kV, 240-MVA, three-phase auto transformer design. The streaming currents and charge densities generated throughout the hydraulic oil flow loop are calculated. The model predicts the dependence of streaming current on oil temperature and velocity. The effect of space charge accumulation in the upper plenum region of the transformer on the electric field under a-c energization is also modelled. Nontrivial enhancements in the electric field at coil edges in the upper plenum are predicted. This result may help to explain the failure mode of recent transformer failures that involved flashovers across large oil distances of tens of inches at the top of conservator-equipped, shell form transformer windings [13].

Streaming Current Theory

The transformer model calculations to be performed in this paper assume that the exit streaming current generated by oil flow electrification in insulating or metallic ducts is given by

$$I = I_\infty(1 - \gamma) + I_o\gamma \tag{1}$$

where γ is the decay factor

$$\gamma = e^{-Z_o/V\tau} \tag{2}$$

In Eq 1, I_o is the current which enters the duct and I_∞ is the limiting streaming current which is a function of duct geometry, oil velocity, oil temperature, and the adsorption rate of charge at the duct surface. In Eq 2, Z_o is the duct length, V the average oil velocity, and τ is the characteristic decay time for charges in the oil given by

$$\tau = \epsilon\epsilon_o/\sigma \tag{3}$$

where ϵ is the relative dielectric constant for oil and σ is the oil electrical conductivity.

According to Eq 1, the exit streaming current approaches the limiting current I_∞ for very long ducts, and any current I_o entering decays exponentially along the duct length.

The theoretical derivation of the limiting streaming current I_∞ is developed in the Appendix and is defined by Eqs A15 and A18 as

$$I_\infty = fWV(\epsilon\epsilon_o RT/ndZF)(\delta/\delta_B)^m \tag{4}$$

where the factor f has been introduced to account for enhanced charge separation due to surface roughness and protrusions such as spacer blocks in the insulation construction. The parameters in Eq 4 are:

f = surface roughness factor, 2 for metal and 5 for insulation ducts,
W = width of rectangular ducts, $W = \pi r$ for pipe of radius r,
V = average flow velocity,
ϵ = dielectric constant, 2.26 for oil,
ϵ_o = permittivity of free space, 8.85×10^{-12} F/m,
R = gas constant, 8.314 J/mole-°K,

T = absolute temperature,
n = transference number of ions $0 < n < 1$, assumed $n = 1$,
d = diffusion layer thickness at surface, Eq A16 in Appendix,
Z = unit charges per ion, assumed $Z = 1$,
F = Faraday number, 96 500 C/mole,
δ = Debye length, $\delta = \sqrt{D\tau}$ where $D = 10^{-5}$ cm^2/s is assumed,
δ_B = boundary layer subzone thickness, Eq A19 in Appendix, and
m = empirical parameter, $m = 0.3$ to 1.0.

Equation 4 calculates I_∞ at a fixed temperature, oil flow rate, and duct configuration. The theory is, in general, valid in the turbulent flow regime and is expected to be less accurate for laminar flow. The limiting streaming current given by Eq 4 can be adjusted to match experimental data over a limited temperature range. Over a wider temperature range the temperature dependence $I_\infty(T)$ is estimated by the empirical expression

$$I_\infty(T) = I_\infty(\sigma_o/\sigma)^a \tag{5}$$

where I_∞ is given by Eq 4, σ_o is an adjustable conductivity, σ is the conductivity at temperature T, and a is an adjustable empirically determined parameter near unity. The explicit temperature dependence on streaming current predicted by Eq 5 may be written as

$$I_\infty(T) = I_\infty(REF)(T/T_o)(\nu_o/\nu)^{(7m+5)/8}(\sigma_o/\sigma)^{(2a+m)/2} \tag{6}$$

where $I_\infty(REF)$ is calculated by Eq 4 at some set of reference values T_o, ν_o, σ_o of the temperature dependent parameters T, ν, and σ where $\nu(T)$ and $\sigma(T)$ are known functions of temperature, and ν is the kinematic viscosity.

In a number of flow electrification model tests performed by Westinghouse, Eq 5 was found to correlate with test data for values of the adjustable parameters $m = 0.3$ and $a = 1$. With these values for m and a, $I_\infty(T)$ is predicted by Eq 5 to vary as $(\nu^{-0.8875} \times \sigma^{-1.15} \times T)$ or approximately as $T/\nu\sigma$. The conductivity $\sigma(T)$ is an increasing function of T while $\nu(T)$ is a decreasing function of T. The combined temperature dependence is found to result in a peak in the streaming electrification for transformer oil at some critical temperature T^*.

Shell Form Transformer Model for Flow Electrification Analysis

A schematic view of a three-phase, shell-form transformer is shown in Fig. 1. The transformer is divided into twelve sections, A through L, which comprise the hydraulic flow loop. Due to the oil flow, charges are separated at the oil-paper interfaces in Sections C and D in the pancake coils and insulating structure (leakage ducts) and at the oil-metal interfaces in the cooling loop Sections H through K. The pumps L also contribute to charge separation. This flow electrification process results in a net accumulation of charge in the bulk oil in the Sections A, B and F, G below and above the coils where the oil velocity is low. The charge density distribution developed in these plenum sections is important in understanding potential failure modes. As charge accumulates in the bulk oil, charge of opposite sign accumulates on insulation surfaces, such as pressboard and spacer blocks. The distribution of surface charge throughout the insulation structure results in the development of high local electrostatic voltage gradients which may be sufficient to trigger spontaneous surface discharges. In most transformer cases the bulk oil is charged positively and the insulation surfaces charged negatively.

The core is represented as Section E. It is assumed that the core plays no role in the electrification process. However, the model does account for the fraction of the total oil flow volume which is lost to the core.

Under steady state oil flow and temperature conditions, the sum of streaming currents enter-

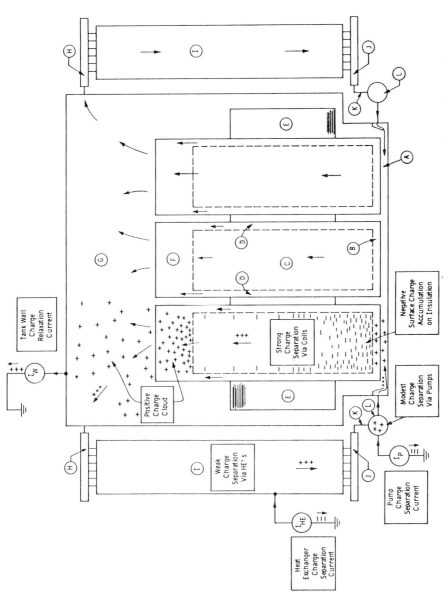

FIG. 1—Model of three-phase transformer for flow electrification analysis. KEY: A = lower plenum; B = insulation entrance; C = coils; D = leakage ducts; E = core; F = insulation exit, bridge; G = upper plenum; H = pipe to HE; I = heat exchanger (HE); J = pipe from HE; K = pipe to pump; L = pump.

ing each of the Sections A through L is equal to the sum of the streaming currents exiting in each of these sections. By applying the streaming current analysis developed above to every unit duct in each section of the transformer, the streaming current inventory and balance may be determined. Equations 1 through 4 are used to compute the streaming currents exiting each section by summing the contributions from individual ducts in a given section. The current balance is achieved when

$$\sum_{j=A}^{L} I_o(j) = \sum_{j=A}^{L} I(j) \tag{7}$$

that is, when the sum of entrance currents equals the sum of exit currents. The charge density at the exit of the j-th section is the ratio of the total exit streaming current and the volumetric flow rate for that section, namely

$$\rho(j) = I(j)/Q(j) \tag{8}$$

The exit streaming current $I(j)$ is

$$I(j) = \sum_{i=1}^{N} I(i,j) \tag{9}$$

where $I(i,j)$ is the streaming current for the i-th unit duct in the j-th section and N is the total number of unit ducts. $I(i,j)$ is calculated by Eq 1 for each duct in the hydraulic oil flow loop.

Streaming Current Analysis for a 240-MVA Transformer Design

The transformer model in Fig. 1 was applied to a 345-kV, 240-MVA three-phase auto transformer design. The results of the static charge analysis are presented in Table 1. The calculated entrance and exit streaming currents I_o and I, and the exit charge density ρ_o for each section A through L of the transformer are given. The % flow, oil velocity, number of oil ducts, and the Reynold's number for each section are also given. The calculation assumes a temperature of 30°C, a conductivity of $3 \times 10^{-14} \ \Omega^{-1} \ cm^{-1}$, and a viscosity of 0.16 cm²/s. The total flow rate with all eight of the transformer's 800 gallons per minute (GPM) pumps running is 14.3 ft³/s (0.405 m³/s).

In arriving at Table 1 the following assumptions are made. The entrance and exit Sections B and F are assumed to play no role in the streaming current, that is, these sections neither generate nor relax charge. The total streaming current generated by the eight pumps is assumed to be 0.82 μA or 102.5 nA/pump. Finally, the relaxation of charge in the plenum Sections A and G is given by the factor $1/(1 + v/Q\tau)$, where v is the total volume of the plenum, Q is the total volumetric flow rate, and τ the characteristic relaxation time.

Based upon the results in Table 1, several conclusions are reached for the 240-MVA design. First, the heat exchanger Section I is a good sink, that is, relaxer for charge. Second, the major generators of charge are the coil structure and leakage ducts, Sections C and D. Third, charge densities as high as 6.7 μC/m³ are generated above coil sections. Fourth, at the insulation exit Section F, the total leakage current is 1.59 μA.

The results in Table 1 are for a fixed oil temperature and flow rate. The dependence of leakage current and charge density for the 240-MVA transformer at the coil structure exit, Region C, as a function of oil velocity or number of oil pumps is presented in Figs. 2 and 3. For these calculations a temperature of 40°C, a conductivity of $3 \times 10^{-14} \ \Omega^{-1} \ cm^{-1}$ and viscosity of 0.12 cm²/s are assumed. The theory predicts a power dependence of leakage current on oil velocity

TABLE 1—*Summary of static charge analysis for 240-MVA transformer.*

Region of Transformer	Flow, %	Effective Velocity, ft/s[a]	Number of Ducts	Reynolds Number	I_o, µA	I, µA	ρ, µC/m³
A. Lower plenum	100	0.10	1	. . .	0.90	0.61	1.50
B. Insulation entrance	95	0.12	3	. . .	0.61	0.61	1.50
C. Coil structure	80	2.00	384	494	0.48	1.18	3.64
D. Coil insulation	15	1.53	60	191	0.092	0.41	6.70
E. Core	5	. . .	. . .	. . .	0.031	0.031	1.52
F. Insulation exit	95	0.12	3	. . .	1.59	1.59	4.13
G. Upper plenum	100	0.07	1	. . .	1.62	0.13	0.32
H. Pipes to HE	100	15.14	4	36 000	0.13	0.15	0.36
I. Heat exchanger	100	0.20	6144	97	0.15	0.013	0.03
J. Pipes from HE	100	15.14	4	36 600	0.013	0.085	0.21
K. Pipes to pumps	100	9.10	8	26 400	0.085	0.080	0.20
L. Pumps	100	. . .	. . .	. . .	0.080	0.90	2.22

NOTE: HE = heat exchanger.
[a]1 ft/s = 0.3048 m/s.

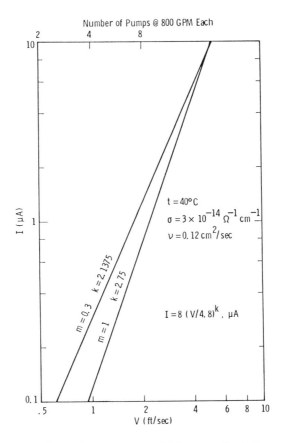

FIG. 2—*Leakage current at coil exit versus oil velocity.*

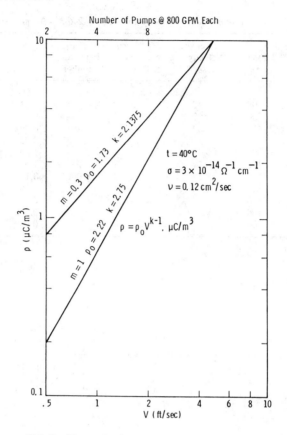

FIG. 3—*Charge density at coil exit versus oil velocity.*

between 2.1375 and 2.75 for values of the adjustable parameter $m = 0.3$ and 1, respectively. The calculations in Table 1 were generated for a value of $m = 0.3$. The charge density dependence on oil velocity is shown in Fig. 3. The predicted charge density increases with oil velocity to the 1.1375 power for $m = 0.3$ and to the 1.75 power for $m = 1$.

Temperature Dependence of Streaming Current

The calculations presented thus far assume a fixed oil temperature. Since the oil temperature in operational transformers is variable, depending on electrical load and environmental conditions, it is important to assess the dependence of the flow electrification process on temperature. Equation 6 was used to predict the temperature behavior of the leakage current for the 240-MVA transformer at the coil exit, Section C. For this analysis, the oil velocity in the coil ducts was assumed to be 2 ft/s (61 cm/s) as in Table 1. The temperature dependence on oil conductivity is approximated by the expression

$$\sigma = 0.30 \times 10^{t/40} \ (10^{-14} \ \Omega^{-1} \ cm^{-1})$$

where t is °C. The oil viscosity as a function of t is approximated by

$$\nu = 0.76 \times 10^{-t/50} \ (cm^2/s)$$

for 0 to 40° and

$$\nu = 0.30 \times 10^{-t/100} \ (\text{cm}^2/\text{s})$$

for 40 to 100°. The results are presented in Fig. 4. The two curves are for values of the parameter $m = 0.3$ and $m = 1$ with the parameter $a = 1$ in both cases. The theory predicts a peak in the coil exit streaming current between 40 and 50°C at 2 ft/s (61 cm/s) oil velocity, that is, all eight pumps running.

Power Frequency Voltage Effect on Streaming Electrification

The transformer model results presented above are for an unenergized transformer. When the transformer is energized the generation of streaming current increases. Large experimental model tests performed by Westinghouse on shell form insulation structures have shown that streaming current development in windings is increased 1.5 to 2.0 times with the application of 60-Hz voltage at practical applied stress levels. Since the effects of a-c voltage on streaming current are not taken into account in the theory presented here, the streaming current and charge density results presented in Table 1 for the 240-MVA transformer are expected to be greater when the transformer is energized. The applied power frequency voltage will lead, there-

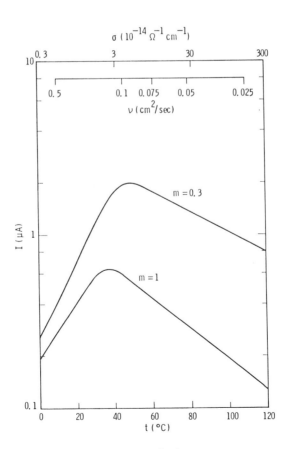

FIG. 4—*Leakage current at coil exit versus temperature.*

fore, to more accumulation of surface charge on insulation and increased space charge in the lower and upper plenums. The true electrical stress distribution in the transformer is determined by the superposition of flow electrification fields on the fields generated by the power frequency voltage and also by transient voltages impressed at terminals.

Theoretical treatments of the influence of a-c energization on streaming current have been developed by Tanaka [19] and by Melcher [20] for oil-paper configurations. These theories show that under a-c energization the oscillations of ions distributed in the oil phase near the oil-paper interface due to the impressed field lead to enhanced streaming currents. At a fixed applied stress and oil temperature, the streaming current enhancement varies inversely with the applied frequency.

Charge Density Distribution in Upper Plenum

As the streaming current emerges from the top of the coil structure, charge accumulates in the bulk oil in the upper plenum. Under energized conditions this "cloud" of space charge which develops in the plenum will distort the normal a-c electrostatic field and potential distributions. In order to assess the impact of space charge on the applied a-c field and potential, the profile of the charge in the plenum is required. Based upon a one-dimensional theoretical model of the upper plenum as shown in Fig. 5, the charge density profile is found to fall off exponentially from the coil exit value of ρ_o to near zero at the tank wall. The a-c potential is assumed to be V_o at the coil exit and zero at the tank wall. An approximation for the variation of the charge density ρ from the tank cover to the coil edge is assumed to be

$$\rho = \rho_o \sinh(\beta x/L)/\sinh(\beta) \tag{12}$$

where the parameter $\beta = L/u\tau$. The distance between the coil exit and the tank cover is L, the average oil velocity in the plenum is u, and τ is the characteristic decay time in the oil. The charge profile predicted by Eq 12 is shown in Fig. 6 for the 240-MVA transformer. In this calculation $u = 0.1$ ft/s (3 cm/s), $\tau = 20$ s, $L = 85$ in. (2.2 m) and $\rho_o = 4 \mu C/m^3$. As shown in Fig. 6, the charge density decays gradually to zero at the wall under these assumptions. At the bridge exit, that is, 25 in. (63.5 cm) above the coil, the charge density falls to 1.3 $\mu C/m^3$.

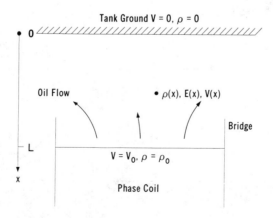

FIG. 5—*One-dimensional model of upper plenum.*

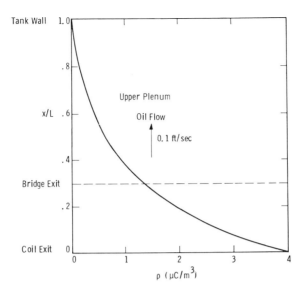

FIG. 6—*Charge density profile above coil.*

Field and Potential Distributions in Upper Plenum

The effect of the space charge distribution shown in Fig. 6 on the electric field E and potential V is found by solving the one-dimensional equations

$$\frac{dE}{dx} = \rho/\epsilon\epsilon_o \tag{13}$$

and

$$E = -dV/dx \tag{14}$$

with ρ given by Eq 12. The boundary conditions are $V = 0$ at $x = 0$ and $V = V_o$ at $x = L$. The results are

$$V/V_o = 1 - x/L \pm K_o(\rho/\rho_o - x/L) \tag{15}$$

and

$$E/E_{AV} = 1 \pm K_o [1 - \beta\cosh(\beta x/L)/\sinh(\beta)] \tag{16}$$

where $E_{AV} = V_o/L$. The parameter K_o is given by

$$K_o = \rho_o L^2/\beta^2\epsilon\epsilon_o V_o \tag{17}$$

The ratio E/E_{AV} given by Eq 16 is plotted in Fig. 7 for the upper plenum region of the 240-MVA transformer. The ratio V/V_o given by Eq 15 is plotted in Fig. 8. The following parameter values were used in these calculations: $L = 85$ in. (2.2 m), $u = 0.1$ ft/s (3 cm/s), $\tau = 20$ s,

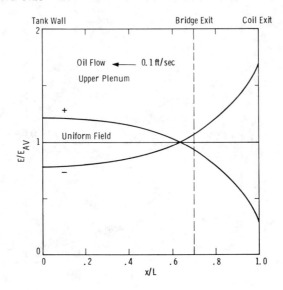

FIG. 7—*Effect of charge density on electric field above coil.*

$V_o = 345 \sqrt{2}/\sqrt{3}$ kV, and $\rho_o = 6.7 \, \mu C/m^3$, the maximum charge density at the coil exit, Region D from Table 1.

As can be seen by Figs. 7 and 8, the influence of the space charge in this case is nontrivial. Depending on the sign of the space charge relative to the sign of the a-c potential, the electric stress at the coil edge relative to the average stress is alternately enhanced and depressed by about ±65% at each half cycle of applied stress. Such high stress enhancements at high voltage coil edges, if actually realized in practice, might help to explain some field failures where leads and coil edges in the upper plenum were involved. It should be pointed out that the space charge

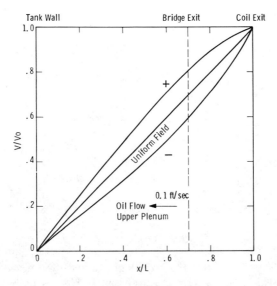

FIG. 8—*Effect of charge density on potential above coil.*

in the upper plenum would also result in stress enhancement on leads crossing above the coil in the bridge region. A more accurate analysis of the complex region of coil edges and leads in the upper plenum is needed to fully assess potential failure mechanisms due to static charge "clouds."

Conclusions

1. Charge separation and thus streaming currents occur in any oil/paper transformer insulation system regardless of transformer construction, that is, shell form or core form.

2. Streaming electrification is a complex multivariable problem. Variables influencing streaming electrification include oil velocity, oil temperature, oil chemistry, duct geometry, solid surface roughness, oil moisture content, contaminants in the oil, and flow turbulence.

3. Streaming electrification in power transformers may enhance electrical stresses within the structure to the point of initiating partial discharges.

4. An engineering model may be used to predict magnitudes of streaming currents and the ensuing electrical stress enhancements.

5. Additional work for other transformer configurations is necessary for completeness and is underway.

APPENDIX

Theory of Flow Electrification

Derivation of Streaming Current for Rectangular Duct

Consider the two-dimensional flow of a low conductivity fluid, such as transformer oil between two parallel surfaces (conducting or insulating) which forms a rectangular duct $2L \times W$ (see Fig. A1).

The equation describing the steady state charge distribution $\rho(x,z)$ between the surfaces is

$$-DV^2\rho + \vec{v} \cdot \nabla\rho + \frac{\rho}{\tau} = 0 \qquad (A1)$$

where $\tau = \epsilon\epsilon_o/\sigma$ is the characteristic charge relaxation in the fluid with oil conductivity σ, and D is the diffusion constant.

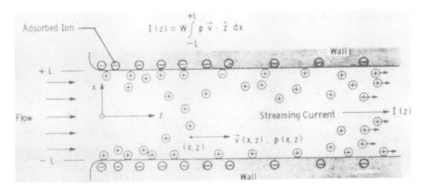

FIG. A1—*Two-dimensional flow electrification.*

In general Eq A1 is difficult to solve for realistic fluid velocity profiles $\vec{v}$. A useful solution is obtained, however, assuming $\vec{v} = u\hat{z}$, that is a uniform velocity profile or plug flow where $\hat{z}$ is a unit vector along the z-axis. The uniform velocity u is an effective velocity defined such that the true streaming current in the fluid per unit distance in the y-direction is

$$I/W = \int_{-L}^{L} \rho\vec{v} \cdot dx = u \int_{-L}^{L} \rho dx \tag{A2}$$

where ρ is the charge density given by the solution to Eq A1 for $\vec{v} = u\hat{z}$. The parameter W is the width of the rectangular duct $2L \times W$ and it is assumed that $W \gg L$.

Assume that the fluid enters the duct with a uniform charge density ρ_o. Furthermore, assume that the first derivative of the charge density ρ at the fluid-surface interface is a parameter α which is independent of the coordinates x and z. These boundary conditions are

$$\rho(x,o) = \rho_o V/u \tag{A3}$$

and

$$\left. \partial\rho/\partial x \right|_{x=\pm L} = \pm\alpha \tag{A4}$$

where V is the average flow velocity in the duct.

Following the treatments of flow electrification by Gavis and Koszman [14–16] and of electrochemical processes by Glasstone, Laidler and Eyring [21], the parameter α is written as

$$\alpha = \pm\frac{ZFC_o}{nd}(1 + D/ndk_A)^{-1} \tag{A5}$$

where

Z = the number of unit charges per ion,
F = Faraday's number,
C_o = the ionic concentration,
n = the transference number,
d = the diffusion layer thickness, and
k_A = the rate of ionic adsorption at the oil-wall interface.

The $+/-$ sign is chosen for positive/negative streaming current, that is, for negative/positive ion adsorption at the fluid-surface interface. The rate of ionic adsorption of the surface k_A according to the theory of absolute reaction rates is

$$k_A = C_A \frac{kT}{h} e^{(\Delta F - \Delta F*)/RT} \tag{A6}$$

where

C_A = the concentration of adsorption sites on the surface,
ΔF = the free energy change due to external factors such as the application of an external field, and
$\Delta F*$ = the free energy of activation for the adsorption process.

In arriving at Eq A5 the reverse reaction to ionic adsorption, that is, escape of ions from the surface, is considered negligible.

Now given the boundary conditions Eqs A3 and A4, the solution to Eq A1 for ρ can be shown to be

$$\rho(x,z) = \frac{\alpha\delta \cosh(x/\delta)}{\sinh(L/\delta)} + \left(\rho_o - \frac{\alpha\delta^2}{L}\right)\exp\left(-\frac{z}{u\tau}a_o\right)$$

$$- \frac{2\alpha\delta^2}{L}\sum_{j=1}^{\infty}\frac{(-1)}{b_j}\cos(j\pi x/L)\exp\left(-\frac{z}{u\tau}a_j\right) \tag{A7}$$

where $\delta = \sqrt{D\tau}$ is the Debye length. The coefficients a_j and b_j are

$$a_j = \frac{\sqrt{1 + 4b_j(\delta/u\tau)^2} - 1}{2(\delta/u\tau)^2} \tag{A8}$$

$$b_j = 1 + (j\pi\delta/L)^2 \tag{A9}$$

Now for $\delta/u\tau \ll 1$, which is the case in practical transformer situations, Eq A7 can be approximated by

$$\rho(x,z) \simeq \frac{\alpha\delta \cosh(x/\delta)}{\sinh(L/\delta)}(1 - e^{-z/u\tau}) + \rho_o e^{-z/u\tau} \tag{A10}$$

The streaming current exiting from the duct at $z = Z_o$ is found by inserting Eq A10 into Eq A2 and integrating. The result is

$$I \simeq 2Wu\alpha D\tau(1 - e^{-Z_o/u\tau}) + 2WLV\rho_o e^{-Z_o/u\tau} \tag{A11}$$

Limiting Streaming Current

According to Eq A11 the limiting streaming current, that is, for $Z_o \to \infty$, is predicted to be

$$I_\infty = 2Wu\alpha D\tau = \pm 2Wu\tau\frac{ZFDC_o}{nd}(1 + D/ndk_A)^{-1} \tag{A12}$$

The bulk electrical conductivity σ may be expressed in terms of the ionic concentration C_o as

$$\sigma = 2ZFDC_o\,(ZF/RT) \tag{A13}$$

If Eq A13 is inserted in Eq A12, then the limiting streaming current may be expressed as

$$I_\infty = \pm Wu\epsilon\epsilon_o\left(\frac{RT}{ZF}\right)\left(\frac{1}{nd}\right)(1 + D/ndk_A)^{-1} \tag{A14}$$

When the electrification process is diffusion controlled $D/ndk_A \ll 1$ and the limiting streaming current reduces to

$$I_\infty = \pm Wu\epsilon\epsilon_o\left(\frac{RT}{ZF}\right)\left(\frac{1}{nd}\right) \tag{A15}$$

which is the form employed in modelling transformer electrification in this paper. In order to compute I_∞ using Eq A15, the effective velocity u and the diffusion film thickness d must be determined.

The diffusion thickness is assumed to be the same as that for mass transport in the fluid which for pipe flow may be approximated by

$$d = 2r/N_{Nu} \tag{A16}$$

where r is the pipe radius and N_{Nu} the Nusselt number. The Nusselt number can be approximated for high Schmidt number fluids by [15,16]

$$N_{Nu} = 0.0223 N_{Re}^{7/8} N_{Sc}^{1/4} \tag{A17}$$

where

$N_{Sc} = \nu/D$ is the Schmidt number and
$N_{Re} = 2rV/\nu$ is the pipe Reynolds number.

The effective velocity u depends upon the Debye length and boundary layer thickness. An empirical estimate of this dependence is expressed as

$$u = V(\delta/\delta_B)^m \tag{A18}$$

where

$\delta/\delta_B = $ the ratio of the Debye length to the boundary layer thickness, and
$m = $ an adjustable parameter near unity.

It is assumed that δ_B can be approximated by the laminar subzone boundary layer which for pipe flow is

$$\delta_B = 50r/N_{Re}^{7/8} \tag{A19}$$

For a rectangular duct where $W \gg L$, the pipe radius r in Eqs A16 and A19 is replaced by L.

References

[1] Klinkenberg, A. and Van der Minne, J. L., *Electrostatics in the Petroleum Industry,* Elsevier Publishing Co., Amsterdam, 1958.
[2] Shafer, M. R., Baker, D. W., and Benson, K. R., *Journal of the National Bureau of Standards,* Vol. 96C, No. 4, 1965, pp. 307–317.
[3] Itoh, M. et al., *Transactions of the Institute Electronic Engineers of Japan,* Vol. 93-A, No. 5, 1973, pp. 175–183.
[4] Kan, H. et al., *Mitsubishi Denki Giho,* Vol. 52, No. 12, 1978, pp. 915–919.
[5] Shimizu, S., *IEEE Transactions on Power Apparatus and Systems,* Vol. PAS-98, No. 4, 1979, pp. 1244–1250.
[6] Higaki, M. et al., *IEEE Transactions on Power Apparatus and Systems,* Vol. PAS-98, No. 4, 1979, pp. 1275–1282.
[7] Okubo, H. et al., "Charging Tendency Measurement for Transformer Oil," Transaction A79051-4, IEEE Winter Power Meeting, New York, NY, February 1979, Institute of Electrical and Electronics Engineers, New York.
[8] Tanaka, T. et al., *IEEE Transactions on Power Apparatus and Systems,* Vol. PAS-99, No. 3, 1980.
[9] Tamura, R. et al., *IEEE Transactions on Power Apparatus and Systems,* Vol. PAS-99, No. 1, 1980, pp. 335–343.
[10] Takagi, T. et al., "Reliability Improvement of 500 kV Large Power Transformers," CIGRE Paper, 12-02, 1978.
[11] Oommen, T. V. and Petrie, E. M., *IEEE Transactions on Power Apparatus and Systems,* PAS-103, No. 7, 1984, pp. 1923–31.
[12] Oommen, T. V., "Static Electrification Properties of Transformer Oil," IEEE 1986 Annual Report, Conference on Electrical Insulation and Dielectric Phenomena, 86CH2315-0, Institute of Electrical and Electronics Engineers, New York, 1986, pp. 206–213.
[13] Crofts, D. W., "The Static Electrification Phenomena in Power Transformers," IEEE 1986 Annual

Report, Conference on Electrical Insulation and Dielectric Phenomena, 86CH2315-0, Institute of Electrical and Electronics Engineers, New York, 1986, pp. 222–236.

[14] Gavis, J. and Koszman, I., *Journal of Colloid Science,* Vol. 16, 1961, pp. 375–391.

[15] Koszman, I. and Gavis, J., *Chemical Engineering Science,* Vol. 17, 1962, pp. 1013–1040.

[16] Gavis, J., *Chemical Engineering Science,* Vol. 19, 1964, pp. 237–252.

[17] Schön, G., *Elektrostatische Aufladungsvorgange und ihre Zundungsgefahren,* Handbuch der Raumexplosion, Verlag Chemie, 1965, pp. 302–366.

[18] Abebian, B. and Sonin, A. A., *Journal of Fluid Mechanics,* Vol. 120, 1982, pp. 199–217.

[19] Tanaka, T., Yamada, N., and Yasojima, Y., *Journal of Electrostatics,* Vol. 17, 1985, pp. 215–234.

[20] Melcher, J., Lyon, D., and Zahn, M., "Flow Electrification in Transformer Oil/Cellulosic Systems," IEEE 1986 Annual Report, Conference on Electrical Insulation and Dielectric Phenomena, 86CH2315-0, Institute for Electrical and Electronic Engineers, New York, 1986, pp. 257–265.

[21] Glasstone, S., Laidler, K. J., and Eyring, H., *The Theory of Rate Processes,* Chapter X, McGraw Hill, NY, 1941.

Dan W. Crofts[1]

Static Electrification in Power Transformers

REFERENCE: Crofts, D. W., **"Static Electrification in Power Transformers,"** *Electrical Insulating Oils, STP 998*, H. G. Erdman, Ed., American Society for Testing and Materials, Philadelphia, 1988, pp. 136–151.

ABSTRACT: The subject of static electrification and related transformer failures has gained the attention of the industry in the last decade. To some extent electrification activity occurs with flowing insulating oil passing over insulation items within every transformer. The extent of the electrification determines whether or not this activity is detrimental to the dielectric integrity of the transformer.

Within the paper, failures are described—two of which occurred on the TU Electric system. Several others have occurred in the United States and even more worldwide. Many of the actual failures have not been well documented and statistical information is lacking in the literature.

Electrification activity is enhanced by many variables which are described in the text. Methods of minimizing detrimental activity are discussed and the need for monitoring techniques is stressed.

KEY WORDS: static electrification, electrostatic charging tendency, streaming electrification

Introduction

With a growing list of failures of large power transformers attributed to the static electrification phenomena, many affiliated with the electric power industry are concerned with the search to find an accurate understanding of the phenomena. Methods to monitor and minimize static electrification activity are needed to prevent additional destructive events.

The paper briefly describes two failures that occurred on the TU Electric system and how these failures relate to static electrification. In addition, an update will be given of the known failures, which involves several manufacturers of transformers and several suppliers of insulating oil.

Also covered are transformer characteristics which enhance static electrification development, including design, insulating fluid, insulation material, other construction material, contamination, and operation of transformer ancillary systems.

Since the oil and the transformer externals are easily accessible, attention to better monitoring methods are essential for risk assessment and management. The characteristics of the insulating fluid can serve to provide an indicator of internal distress in a transformer, even though the fluid itself may not be the sole contributor.

Failures

Japanese—In 1972, there were two (assumed from comments made at the 1986 Conference on Electrical Insulation and Dielectrics Phenomena [22]) large power transformer failures while undergoing tests at the factory in Japan. Both units had voltage ratings of 500 kV with power ratings of 750 and 1000 MVA, respectively. One unit was of a shell-form design and one was of a core-form design. Both failures were attributed to streaming electrification. There also is a high probability of field failures not reported to the industry.

[1]Manager of Transmission and Distribution Operations, TU Electric, Dallas, TX 75266-0268.

CIGRE—Publications of the International Conference on Large High Voltage Electric Systems (CIGRE) have mentioned static electrification problems [2].

Consolidated Edison—In January 1979, an experimental compact d-c terminal constructed, operated, and studied at the Astoria station on the Consolidated Edison Co. system encountered a static electrification problem in the converter valve cooling system. Cooling for the converter valves is accomplished by circulating Refrigerant 113 (trichlorotrifluoroethane) through fiberglass epoxy–reinforced tubes and insulating Tefzel (modified copolymer of ethylene and polytetrafluoroethylene) tubes. Circulation of this refrigerant through these tubes resulted in spark discharges that eventually punctured the tubing and resulted in coolant leaks.

In a solution to the problem, an antistatic additive (Mobil DCA-48) was added to the coolant. To date, there has been no recurrence of the static electrification problem. This terminal has seen only intermittent operation and has never been commissioned by Consolidated Edison [9,20,26].

Public Service Electric and Gas (PSE&G)—A unit manufactured for PSE&G experienced internal flashovers from static electrification while undergoing tests at the manufacturer's factory. When the cooling pumps were run, discharges could be heard and were even seen by factory personnel. Other identical units at the factory did not display the same phenomena. The only difference between the transformers was the oil in the troubled unit. The oil was from a different refinery than that used in the other transformers.

TU Electric—On 12 Dec. 1983, TU Electric experienced a failure of a three-phase, 345/138-kV, 450-MVA, shell-form, conservator-equipped autotransformer at Renner Substation. The failure occurred approximately three months after carrying load for the first time, with no forewarning of distress [11–13,16].

The failure involved a flashover in the main tank from the "C" phase high-voltage winding near the line end to a low-voltage lead going to the no-load tap changer near the neutral end (see Fig. 1). The flashover traversed approximately 42 cm (16.5 in.) of oil space. The failure was severe and resulted in an oil spill of approximately 79.5 m^3 (21 000 gal).

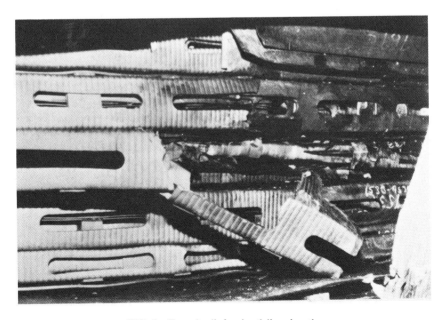

FIG. 1—*Top of coil showing failure location.*

The unit was returned to the manufacturer for teardown, inspection, and rebuilding. During the latter part of March of 1984, manufacturer's personnel, in conjunction with TU Electric representatives, began the layer-by-layer disassembly of the damaged winding phase packages. We were searching for an underlying reason for the flashover of 42 cm (16.5 in.) of oil above the "C" phase winding. Layer-by-layer disassembly of the unit disclosed *many* areas of potential electrification failure. There was evidence of tracking apparently caused by static electrification on coils, static plates, insulation pieces, and oil ducts throughout the transformer. This tracking followed oil circulation paths. There was also evidence of carbon deposits, but closer inspection revealed evidence of heavier deposits in areas of higher oil velocity (the carbon particles in the oil apparently assume a positive charge and seek areas of negative potential).

In addition, low energy electrical punctures in the conductor insulation were found in a coil in Phase "C." These could have eventually led to failure because the punctures were bridged by a carbonized path in adjacent insulation pieces to other conductors in the same winding layer. The activity was turn to turn. There was no evidence of a-c power follow current between the conductors.

Figure 2 shows another type of tracking, which we have erroneously termed as "wormholes," found in insulation washers throughout the unit. These "wormholes" are subsurface paths presumably caused by static electrification (apparently negative charges seeking a path to ground within the pressboard). Studies by the manufacturer have discounted moisture or contaminants within the pressboard as causative. Nevertheless, notice that these electrification paths did follow the oil flow. Throughout the transformer, some of these subsurface paths were carbonized while others are not. Conversely to our manufacturer's opinion, the Japanese authors, during the 1986 Conference on Electrical Insulation and Dielectrics Phenomena (CEIDP), offered the theory that "wormholes" in the insulation are created when moisture within the pressboard expands from the heat generated from static electrification leakage current at the surface. That, in effect, causes bubbles to form within the insulation creating pressure with the obvious consequences. The credence of this hypothesis remains to be determined.

Figure 3 shows evidence of static electrification on the bottom of "B" phase at the joint closure. This photograph was taken at the bottom end of the coil package. Also, the affected areas are where oil exits from pumps into the lower plenum. Vorticity patterns and possible turbulence are existent, yet additional velocity and vorticity changes occur as the oil enters the windings in the joint closure area. The disassembly of this phase and Phase "A" disclosed similar evidence of static electrification throughout the phase packages [11–13,16].

FIG. 2—*Subsurface tracking or "wormholes."*

FIG. 3—*Lower section of winding showing electrification activity at the joint closure.*

The particulars of the failure at Renner parallel many of the contributing factors discussed in the IEEE transactions referenced. The operating temperature of our autotransformer was in the intermediate range (18° to 48°C). The unit was lightly loaded with the cooling running in the manual mode. This unit was equipped with eight pumps rated 4.476 kJ/s (6 hp) and 3.03 m³/s (800 gal/min). The moisture content of this unit was 7 ppm.

The unit contained transformer oil displaying peculiar charging tendencies. Charging tendency tests were made on the failed unit's insulating oil after the failure. As a result of the failure, the oil samples were contaminated with carbon and other impurities. The charging tendency of the oil prior to failure was estimated to be in the 600 to 700 μC/m³ range from tests made following failure.

On 24 May 1984, a second 450-MVA autotransformer, identical to the Renner unit (built on the same shop order), failed in service at the TU Electric Anna Switching Station. An internal inspection revealed damage similar to the Renner unit [16].

The failure again involved a flashover in the main tank from the "C" phase high-voltage winding near the line end to the low-voltage lead going to the no-load tap changer near the neutral end. The flashover again traversed approximately 42 cm (16.5 in.) of oil space at the top of the winding. Another oil spill occurred as 79.5 m³ (21 000 gal) of flaming oil spilled from the transformer. It was obvious that the transformer had been on fire and had self-extinguished. The internal inspection also revealed a flash across the bottom porcelain of the "B" phase high-voltage bushing. This was unlike the Renner failure. However, oscillograph records indicated the fault started on "C" Phase. Carbon deposits and fire damage to the exposed top section of the shell-form transformer windings made it impossible to find evidence of static electrification. Prior to failure all eight pumps were running (mistakenly left in the manual "ON" position) with the oil temperature in the intermediate [313 to 318°K (40 to 45°C)] range. The pumps had been on for only twelve days. The oil had a charging tendency measurement of 970 μC/m³ prior to failure and was from the same oil supplier as in the previous failure. In both cases, the oil was furnished by the transformer manufacturer in accordance with their specifications.

In an attempt to prove or disprove the static electrification theory as a failure cause, a pump was removed in an attempt to see the bottom of the winding through the pump mounting area. This did not work; however, a hole approximately 61 by 38 cm (24 by 15 in.) was cut in the side of the transformer at the pump mounting area. With mirrors, electronic flash, and a telephoto lens, pictures of the bottom of the "C" phase coil package were taken. These pictures showed the same type of tracking as was seen on the previous failed transformer.

The failed Anna unit was returned to the manufacturer for untanking and investigation. A layer-by-layer inspection of the "C" phase winding showed evidence throughout of static electrification, apparently once again the cause of the failure. There was tracking on coils, insulating washers, oil ducts, and joint closures. Numerous locations of incipient or potential failures were seen throughout the unit. Evidence of subsurface paths ("wormholes") in the pressboard and insulation punctures were also found.

The bottom end of "A" and "B" phase packages had tracking similar to "C" phase. The tracking was most evident in the areas where there are changes in oil flow parameters. Also, the tracking was very similar to that found on the bottom end of the phase packages on the Renner transformer. In both the Renner and Anna failures, the static electrification damage was so severe that eventual transformer failure would have occurred exclusive of the flashovers above the windings. This represents several failure modes operating simultaneously.

Union Electric—On 24 June 1984, a 345/138-kV, 560-MVA autotransformer owned by Union Electric Co. failed at the Belleau Transmission Substation. This transformer is a shell-type design with a total of five oil cooling pumps and failed less than one year from energization while carrying 150 MVA of load. All cooling was running in the manual "ON" position. An internal field inspection indicated the failure was initiated between the high-voltage (HV) winding of "A" phase to a test tap lead. This represents an arc path of 24 cm (9.5 in.) through oil above the windings. The oil was tested for static charging tendency and measured 510 μC/m^3. Damage was not extensive, and the unit was repaired in the field and returned to service [14].

Centrais Electricas Do Sul Do Brasil S.A.—*ELECTROSUL*—In February 1985, ELECTRO-SUL suffered the failure of a 200-MVA, 525/230-kV autotransformer. This transformer is of shell form design and failed violently in service after four years of operation. The unit failed between the outer first turn of the HV winding and the tertiary winding connecting lead. In June 1985, an identical transformer failed in service. The unit also failed between the outer part of the HV winding and the tertiary winding connecting lead. ELECTROSUL personnel feel certain that these two failures are due at least partially to static electrification.

Alabama Power—On 10 Sept. 1985, a single-phase, 500/230-kV, 448-MVA autotransformer owned by Alabama Power Co. failed at the South Bessemer Substation. The unit is a shell-type design and failed several hours following energization without load. The pumps were running continuously in the manual "ON" position. Static charging tendency of the oil was 227 μC/m^3. The oil was from a different supplier than that used in the failed TU Electric and Union Electric transformers. These units, however, were built by the same manufacturer. The Alabama unit and the two TU Electric transformers were disassembled in the factory. The manufacturer disclosed the pressboard to be from the same supplier of insulating paper (screen-like surface texture).

Factory inspection of the Alabama unit revealed that the fault was initiated between the "A" phase HV series lead and the "A" phase LV series lead located 33 cm (13 in.) apart above the windings. The normal operating potential between these points is approximately 288-kV. Signs of tracking were also found in the bottom section [18].

Another identical unit in the same three-phase bank at Alabama Power was given electrical field tests since undergoing similar field operating conditions. The unit failed the field test and was returned to the factory where the field test failure was duplicated. Teardown of the unit disclosed a manufacturing defect to be the failure cause. Layer-by-layer disassembly of this unit did *not* show any evidence of static electrification activity. The pressboard insulation, however, was smooth and from a different supplier than other-mentioned failures, as per the manufacturer [27].

American Electric Power (AEP)—In November 1985, AEP experienced a failure of a single-phase, 500-MVA, 765/345-kV autotransformer. This transformer failed during insulation testing. The point of failure was from the high-voltage winding to the neutral lead. In December 1985, an identical transformer failed violently and caught fire after 7.5 h of service at no load. The failure was again from the high-voltage winding to the neutral lead. It is surmised that these two transformers failed due to static electrification.

In January 1986, a single-phase, 500-MVA, 765/26-kV generator step-up transformer failed in service while at minimum load. The failure occurred between a high-voltage series crossover and top leads at the neutral end of the windings. This failure is believed to have been possibly caused by static electrification.

South Africa—In 1986, a South African utility also experienced a failure of a large power transformer of core form design. Unconfirmed reports have indicated design concerns were included in the manufacture in an attempt to prevent static electrification flashovers at the upper winding exits. The failure, however, was located near the bottom. The unit was also equipped with a conservator. Excessive cooling was apparently included in the design of this unit [28]. At the April 1987 Doble Conference, it was inferred that a similar failure of a previous unit was probably due to electrification rather than resonance as previously postulated.

Australia—In 1987, the Electricity Council of New South Wales suffered two large generator step-up transformer failures. These transformers are rated 23/330 kV, 390 MVA and are of the same design and manufacturer. They are of core form design with concentric windings built on five horizontal and two vertical legs. The transformers are outfitted with two oil pumps rated 5.68 m^3/s (1500 gal/min) that are wired for automatic control.

One of the failed units has been untanked and inspected at the factory and found to have an apparent turn-to-turn failure in the center portion of the HV winding of one phase. Inspection did not reveal, however, any evidence of tracking or "wormholes" that are typically found in failures attributed to static electrification. The other failed unit, at the time of this writing, has not been untanked, but the field internal inspection reveals evidence of similar damage.

Tests, including dissolved gas in oil, made previous to the failures did not indicate any abnormalities. Since these two failures, the utility has performed partial discharge tests on identical in-service units, but, for the most part, all of the investigative effort is devoted to the failed units. The utility is investigating all possible failure modes, such as excessive stress distribution and static electrification or a combination of the two.

Other—Several other transformer failures that are unexplained have also been suspected to be a result of static electrification. Unfortunately, the evidence was destroyed long before static electrification modes were well known.

Failure Scenario—The following scenario is offered in an attempt to rationalize how static electrification failures may have occurred at the top of the windings. Electrification failures, however, can occur anywhere within an operating transformer:

For cooling purposes, the forced oil flow in power transformers is directed into the bottom of the windings and out the top. Charged are generated, carried, and distributed by this flow of oil across insulation in the oil path. Through this process, and other methods of charge generation or injection, an impaired dielectric path could be created over which a power discharge arc could occur where the oil exits at the top of the windings. The impaired dielectric path may be only clouds of charges impairing dielectric strength in the oil space above the windings. The power follow current would then cause the failure of the transformer. The previously mentioned impaired dielectric path could also result from d-c potential differences allowing d-c flashovers, a corollary to those occurring in nature during a thunderstorm. Prior researchers of the static electrification phenomena have hypothesized d-c potential differences as high as hundreds of thousands of volts or more. Also, they have actually observed static discharges in power transformers (See Fig. 4).

The South African failures occurred in the bottom of the windings of core form transformers. Perhaps a similar scenario can be developed for these failures [30].

Conditions Influencing Static Electrification

In the late 1970s and early 1980s, the Japanese conducted detailed research of the static electrification phenomena. Much of the research work was published in IEEE transactions [3-8]. Since then, much has been learned from other researchers with many areas not yet clearly understood.

Accordingly, static electrification in power transformers is influenced by several factors including:

1. *Oil temperature.* Charging tendencies of the oil increase with temperature. However, in a transformer, the charging level reaches a maximum due to a charge leakage process referred to as "relaxation." The maximum charging tendency apparently occurs at some intermediate temperature [variable between 303 and 333°K (30 and 60°C)] [1,3-5,7,10]. Current research has not verified that the maximum hypothesis is germane. This may be due to the physical limitations of the research models.

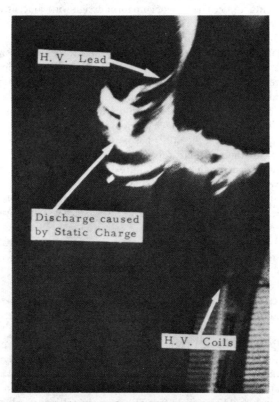

An Example of the Discharge Caused by Streaming Electrification.

FIG. 4—*From Japanese research.*

2. *Oil moisture content.* Charging tendencies of insulating oils increase as moisture levels decrease. Oils with moisture content in the range of 15 ppm or less can have higher charging tendencies [*10*].

3. *Flow rate.* Charging tendencies increase with greater flow rates. Japanese research that displayed the increase varied somewhere between the second and the fourth power of the velocity of the oil flow. Some of the Japanese research was done at high flow rates, higher than most flow rates in transformers manufactured in the United States. For example, the average flow rate involved in our failures was 21.3 cm/s (0.7 ft/s) over an average typical cross section. This flow rate increases to 45.7 to 61.0 cm/s (1.5 to 2.0 ft/s) in coil areas and to greater than 457 cm/s (15 ft/s) near pumps and pipes in the heat exchangers. Ongoing research has not confirmed electrification activity being proportional to as high as the fourth power of the velocity [*24*].

4. *Turbulence of oil flow.* Japanese research indicates: "the charge motion or generation does not depend on diffusion, but turbulent motion of oil" [*5,7*].

5. *Contaminants or surface active agents.* Charging tendencies greatly increase with trace quantities of surfactants such as sulfonates present in the oil. Interfacial tension can be greater than 40 dyne/cm (satisfactory under present standards) with surfactant contamination of less than 50 ppm. This concentration can create undesirable charging tendencies in the oil [*10*].

6. *Charging tendency of oil.* Some oils characteristically have higher charging tendency than others. This could possibly be based on the source of crude, the refining process, and other variables or contaminants yet to be identified. Additional research is needed for establishment of levels of concern for and the importance of charging tendency measurements. We have assumed, however, that ranges for typical new oils may have been in the order of 1 to 150 μC/m^3 (through mid-1986) with less than 25 μC/m^3 for low-charging oils and as high as 1730 μC/m^3 on our highest test for aged oils. The latest tests on new oils have shown that the refiners have apparently addressed the charging tendency. Oils available for use in 1987 appear to test in the order of 25 μC/m^3 or less. As strange as may seem, the refiners have not shared the reason for this apparent decrease.

7. *Surface conditions.* The charge generation/separation process is enhanced with increases in roughness of the solid insulation. The texture of the insulation surface apparently affects charge generation/separation. Insulation materials from different suppliers have different surface characteristics. The pressboard used in our transformers and the Alabama unit involved in failures was apparently from the same supplier, according to the transformer manufacturer. The surface of the pressboard was not smooth (screen-like texture) as a result of the manufacturing process of the paper.

8. *Energization.* Static electrification intensity is proportional to increasing a-c field strength, probably due to charge injection processes. Current research in flowing oil/pressboard systems has shown the magnifying effect of an alternating current field can enhance the localized charge density perhaps as much as five-fold.

9. *Dielectric strength of oil in motion.* The dielectric strength of flowing transformer oil varies with the velocity of flow and is not necessarily the same as when tested at rest. With due respect to model dependency, a sharp dip has been demonstrated at flow rates of near 50 cm/s (1.6 ft/s). Obstructions to oil flow paths and flow direction modifiers can create cellular convection. Likewise, cellular convection has appeared as flow rates leave laminar and approach turbulent velocities. These convection cells have been observed with Reynolds numbers in the 97 to 2500 range. These measurements were made using a particular laboratory model and are dimensionally dependent. There may not be an absolute parallel comparison in a dimensionally different operating transformer. Nevertheless, such convection cells are formed. The range of 97 to 2500 is similar to the Reynolds numbers calculated for flow through the insulating washers of many transformers. Cellular convection in itself apparently enhances the charge injection process into the oil stream [*23,25*]. The dielectric strength of oil also varies with temperature, moisture content, gas bubble presence, contamination, and particulate matter.

10. *Migration of moisture.* In an oil/paper operating transformer system, there is a migration of moisture between the oil and the insulation. As temperature and pressure changes occur in a transformer, the moisture migrates to reestablish equilibrium conditions. As the moisture leaves or enters the surface of the insulation, there is a corresponding change in the conductivity of the fluid at the interface. This condition likewise affects the charge separation process.

11. *Particulate matter.* Particulate matter in the transformer oil from the manufacturing or installation process has an effect on electrification activity in addition to the previously mentioned dielectric characteristics.

12. *Charge injection.* Upstream charge injection has been demonstrated to affect downstream dielectric integrity [29].

13. *Pumps.* Pumps are suspected as being a substantial source of charge generation. This presumption requires confirmation and accurate quantification on an operating transformer or acceptable semblance [29].

14. *Orifices.* Orifice effects have also been demonstrated to generate charge.

Transformer Characteristics Influencing Static Electrification

Not unlike most products, a transformer is designed to operate safely and reliably for a reasonable length of time and be produced at a price that is affordable to the consumer yet profitable to the manufacturer. As technology has advanced, design improvements have been implemented and manufacturing techniques and materials have been improved to effectively produce a product that will meet today's growing power requirements at a minimum cost. Utilities demand larger transformers yet; conversely, shipping limitations impose dimensional and weight restrictions. As an option, many manufacturers pursued improved heat transfer and computer-refined design. In the challenge of meeting these requirements, some factors involving static electrification were neither adequately recognized nor appropriately considered. As the many static electrification variables entered the picture, something finally exceeded the dielectric withstand and failures occurred.

These factors include the design and selection of solid insulation materials, liquid insulation cooling system design, pump design, charge relaxation consideration, application of charge collectors, use of additional conductor clearances, addition of conductivity modifiers, lack of adequate design information, and absence of methods of monitoring.

Insulating Fluid—As mentioned previously, studies have shown that the charging tendency of the liquid insulation is influenced by many variables. Insulating oil standards do *not* adequately address the influence of oil in electrification activity.

A test of transformer oil that is most commonly associated with static electrification is static charging tendency. Liquid insulation possessing a high static charge density has been related to increased potential for static electrification activity. Static charging tendency generally has been measured using the "Mini-Static" testing device (see Fig. 5) adapted by Westinghouse, which has yet to be accepted by industry as the standard test procedure [1]. The true significance of this test remains unresolved, but who can offer a better method of forewarning? Has anyone yet developed an improved method of measuring the accumulation of charges within the oil? Let us concentrate our efforts on formulating a better method rather than criticizing the innovative efforts of others.

Other properties which may identify the liquid insulations charging tendency, such as conductivity and resistivity, are being investigated with favorable optimism.

Why, I ask, have existing oil test methods only considered static oil? Oil in an operating transformer is constantly in motion. Why then haven't standards addressed the characteristics of oil in motion?

Insulating Material—Previous study has shown that the physical properties of the solid insulation material will influence static electrification activity. The type of material used along with its texture and surface patterns will influence the degree of static electrification activity. Generally, the charge separation will be enhanced as the roughness of the solid insulation material increases. The surface characteristics can be greatly influenced by the manufacturing techniques of the insulation items. Unfortunately, however, the removal of insulation for testing purposes is *not even practical* for consideration.

Operation of Transformer Ancillary Systems—Since the liquid insulation in a power transformer also serves as a coolant, it is imperative that a means of efficient heat transfer be designed. This effective heat transfer process usually takes the form of numerous directed cooling ducts, forced oil flow pumps, fans, coolers, and radiators. Historically, cooling systems have been designed to assure that the transformer will operate at a certain temperature rise over ambient temperature for a given load capacity. As a result, transformer cooling systems were designed and built with precarious consideration given to influence on static electrification. Users should operate existing units in a manner to avoid increased exposure to static electrification problems, that is, automatic modes for cooling system operation as was intended in the design.

As mentioned previously, the oil velocity or flow rate, oil temperature, surface conditions, and vorticity of oil flow all influence the existence and degree of static electrification activity. Flow considerations can be controlled by altering transformer design without sacrificing transformer load capacity.

Prudent operation of transformers necessitates understanding of the conditions that minimize static electrification activity. Currently, the state of the art involves operating cooling controls in the automatic mode to avoid susceptible temperature ranges, monitoring of the insulating fluid and modifying if necessary, implementing field modifications as agreed to by the manufacturer, and participating in investigations to better understand static electrification. All of this is being done *without* the existence of a recognized good test or indicator to give forewarning of distress.

Monitoring Methods

Rather than employing "ostrich" techniques and ignoring static electrification failure modes, many have chosen to explore the feasibility of finding a system to provide forewarning of distress. Certain tests are available, but, in themselves, they do not offer adequate monitoring to prevent the "blindsiding" failures that have and will continue to occur.

Insulating Oil Tests

Static Charging Tendency—Currently, the most popular means of identifying transformers exhibiting potential static electrification problems involves the measurement of static charging tendency of the oil (test method not yet approved by ASTM). This test, originally developed for petroleum fuels and later adapted for electrical insulating oils, measures the charge density generated in a volume of oil passed through a filter [4,10]. Some researchers use the term electrostatic charging tendency (ECT). The apparatus used, described in Fig. 5, is an adaptation of the measuring device developed by Exxon in 1972 known as the "mini-static tester" [10].

In principle, an oil sample is forced through a specified filter and a charge separation occurs. The charge on the filter is measured by an electrometer and is usually negative. Charging tendency is defined as the amount of charge generated per unit volume of oil in the flow. The practical unit for charge in the present measurement system is microcoulombs (μC); therefore, charging tendency is usually expressed as $\mu C/m^3$.

TU Electric and companies such as Doble Engineering, Shell, Exxon, Gulf, Westinghouse, and Union Electric built "ministatic" testers and began testing and investigating oils following the failures.

After a period of several years and many tests, factors that affect the measurement of charging tendency have been found. Test parameters such as type of sample bottle, method of sam-

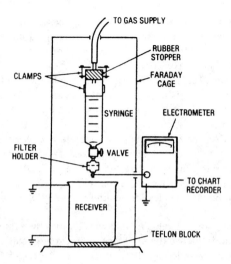

FIG. 5—*Electrostatic charging tendency measurement device.*

pling, effects of storage time, light, exposure, moisture, temperature, agitation, type of filter paper, flow pressures, number of runs, and measurement interpretation are variables needing to be addressed by standards to develop approved methods for consistent results; then maybe guidelines can be established for a tool to monitor electrification activity, perhaps one of several tools.

Oil Diagnostic Tests—Other tests on transformer oil, such as ASTM Test Method for Interfacial Tension of Oil Against Water by the Ring Method (D 971), ASTM Test Method for Water in Insulating Liquids (Karl Fischer Method) (D 1533), ASTM Test Method for Dielectric Breakdown Voltage of Insulating Oils of Petroleum Origin Using VDE Electrodes (D 1816), Percent Power Factor at 100°C, Conductivity (D 3114), and ASTM Test Method for Specific Resistance (Resistivity) of Electrical Insulating Liquids (D 1169-80), have been performed to determine their relation to static charge density. Results of a TU Electric study on in-service oils indicate:

1. A correlation between increasing conductivity (greater than 1.0 pS/m) and decreasing resistivity (less than 10 times 10^{14} Ω-cm) exists with increasing charging tendency of oil (increasing charging tendency is negative). See Figs. 6 and 7.

2. Service-aged oils with high static charging tendency (150 $\mu C/m^3$ or greater) in our tests exhibited high power factor at 375°K (100°C) (greater than 1.0%) and low interfacial tension (less than 40 dyne/cm). The converse did not necessarily hold true.

Dissolved Gas In Oil—Dissolved gas-in-oil analyses from power transformers are excellent tools utility maintenance personnel use to evaluate transformer condition. It is assumed that static discharge activity is very similar to partial discharge activity and will generate a similar gas profile. In terms of percentage of the total combustible gas produced by static discharge activity, hydrogen (H_2) should represent 75% or greater, carbon monoxide (CO) 10 to 15%, methane (CH_4) 5 to 10%, and ethylene (C_2H_4) 0 to 5% [5].

Prior to the Renner failure, a dissolved gas-in-oil analysis was performed 22 days after the unit was energized carrying no load. Results of the dissolved gas-in-oil analysis were:

Carbon dioxide (CO_2)	167 ppm
Hydrogen (H_2)	10 ppm
Carbon monoxide (CO)	7 ppm

CHARGE DENSITY VS. CONDUCTIVITY

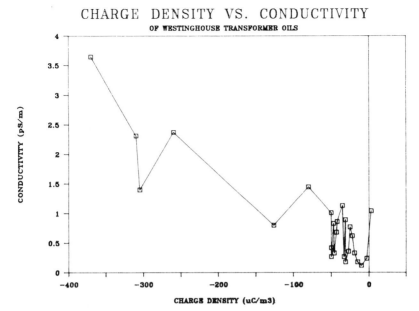

FIG. 6—*Conductivity versus electrostatic charging tendency.*

| Methane (CH$_4$) | 2 ppm |
| Ethylene (C$_2$H$_4$) | 2 ppm |

Due to the low concentration of these gases, little, if any, significance was given to these results before the failure. Even if existing technology hydrogen monitors were installed, these

CHARGE DENSITY VS. VOLUME RESISTIVITY

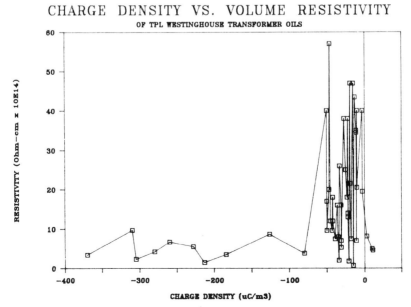

FIG. 7—*Resistivity versus electrostatic charging tendency.*

values would have been ignored and would have been well below the alarm point chosen for the monitors. Since most units normally exhibit higher concentrations of combustibles, it may be difficult to assess how much gas is generated due to static electrification discharges alone. We do, however, continue to monitor each suspected unit for signs of distress by using both dissolved gas in oil tests and installed monitors on suspicious units. Such tests, however, have not detected any conditions anticipating electrification troubles.

External Tests

Acoustical Emission—Partial discharge signals have some similarity to static discharge signals [15]. With this in mind, partial discharge detectors have been used with the intent of identifying static electrification activity. The acoustic emission partial discharge detector (PDD 101) developed by Electrical Power Research Institute (EPRI) under RP 426-1 has been used by TU Electric to detect static discharges. Our experience, thus far, has indicated that higher levels of activity (measured in counts) occur near the transformer pumps. Whether this activity is actually static discharges or possible pump noise remains undecided. Research in this area, such as EPRI RP 426-2, studied the effects of oil pump performance on static electrification and attempted to identify static discharge signals [15]. The results failed to answer the questions.

Electrical Emission—Another instrument designed to measure partial discharges, developed by Biddle Instruments, was used by TU Electric to measure the partial discharge energy in a deenergized transformer. Partial discharge energy measurements were made continuously for 3 h on a deenergized 800-MVA EHV transformer stored at Martin Lake SES. Results from these tests indicated a gradual increase in partial discharge energy when the pumps were turned on and a gradual decrease when the pumps were turned off. This energy was believed related to static electrification. This is further verified by the elevation of charging tendency measurements of the oil following pump operation.

Measurement of Ground Current—After our Renner failure, a Stoddart EMI/Field Intensity Meter (Model 17/27) was used in conjunction with a RF current transformer to measure current activity in the ground connection of several power transformers. Measurements on our transformers with pumps running ranged from a few hundred microvolts to 10 000+ MV. On units that had the pumps mounted directly adjacent to the tanks and with the pumps running, the Stoddart instrument indicated high RF activity on the units with high charging tendency oil. This instrument has been used in conjunction with other identification methods. There is potential application in use of this device together with an acoustical device as inputs to a dual-trace scope to identify static discharges. More work is needed in this area.

Wave Guides—The use of wave guides for acoustical monitoring of partial discharge activity is included in the scope of our ongoing research activities. Experience from application of these devices could prove their effectiveness.

Charging Tendency Reduction Options

As documented in earlier papers, static electrification activity can be minimized or eliminated when the oil flow velocity is reduced by altering pump design or cooling system operation. Transformer retrofill using the original oil treated with Fuller's Earth or using new low charging tendency oil is another effective method of reducing the level of static electrification activity.

Conductivity Modifiers or Charge Suppressant Additives—A possible static charging tendency reduction method needing to be further investigated involves the addition of charge suppressants to the oil. The effectiveness of using charge suppressants, such as BTA (1,2,3-benzotriazol) and AB (alkylbenzene), in transformer oil will require further study. The potential detriments for the use of charge suppressant additives should be further explored. For example, it has been suggested that these additives could produce high-resistance coatings on copper or other construction materials which could lead to tap changer contact problems, etc. Another potential detriment is the suggested possibility that some additives are consumed in the oxidation process and produce water as a by-product. Literature on BTA indicates dielectric strength is reduced somewhat if BTA concentration is greater than or less than approximately 10 ppm [17]. Proposed Institute of Electrical and Electronic Engineers (IEEE) transaction papers disagree with this hypothesis.

Following static electrification flashovers, EPRI RP 1536-7, RP 1536-9, RP 1207, and RP 1291-5 studied the interaction of Mobil DCA48 additive in Refrigerant 113 used as a coolant in the converter valves of a prototype high voltage d-c terminal. Preliminary study demonstrated that a thin organic film was deposited on the electrodes due to electrolysis. This work concluded that the DCA48 addition did not create undue complications in this application and the Refrigerant 113/DCA48 mixture would be an acceptable design parameter for eliminating electrification problems [9, 20, 26].

It may be entirely possible that a contaminant or "gremlin" in the oil may be responsible for the oil's unusual ability to produce static charge. On the other hand, the lack of an important charge suppressant or "guardian angel" in the oil (due to severe refining, etc.) may be responsible as well.

Conclusion

As a doctor examines his patient, so do we monitor and examine a transformer. Much can be learned from the chemistry of body fluids. Likewise, knowledge can be gained from examining the chemistry of transformer fluid.

Some have claimed the transformer oil is pristine, innocent, unquestionable, noncausative to static electrification problems, and not an indicator of static electrification activity. Many others, including myself, must disagree. What if physicians had adopted this philosophy with regard to body fluids? Therefore, rather than ignoring the transformer's oil as a means of potential detection and letting the patient die, why shouldn't the industry consider and pursue monitoring and testing of the fluids as one of the indicators of deleterious electrification activity? Other methods of monitoring may employ systems other than fluid monitoring or testing.

Like it or not, the fluid in a transformer is easily accessible without the necessity of removing an operating unit from service. Therefore, if the fluid is readily accessible for testing, why then has there not been an industry accepted and approved method of testing and monitoring the fluid for electrification activity?

Let us get off dead center and be open minded. We should quit blaming one manufacturer of transformers, one oil refiner, one manufacturer of insulation, one method of operation, etc. *The answer is not that simple!* Static electrification is a highly complex phenomenon that occurs to some extent in *any* transformer. Let's simultaneously give our designers, manufacturers, and refiners the basics they need to design around the deleterious effects while providing the users with the needed tools to prevent catastrophic failures.

When the final truth is known, the problem will involve combinations of designs, operating philosophies, characteristics of oils and insulating materials, and a complex combination of other postulated failure modes. The failure mode may involve bubble evolution, moisture, flow rates, design, resonance or other voltage excursions, preexisting conditions to which the transformer has endured, manufacturing techniques, installation and maintenance procedures, as well as static electrification.

It is doubtful that we will be able to distinguish between some of the causative conditions. Why then doesn't the industry diligently pursue all avenues to define the safe operating and design parameters for all of the failure causes? Once this is done, appropriate monitoring devices can determine, with high degrees of sophistication, when we are operating (or designing) outside of a safe parameter envelope. To do this we must, by necessity, look at all facets including the insulating oil.

In conclusion, let the preceding be considered as input for ASTM consideration. We do *not* have an approved and acceptable test method for insulating oil electrification potential and we *cannot* dissect operating transformers to provide answers. Let us not resort to surgery if other methods are available. Please, ASTM, give us a test method—don't ignore it and let the patient die. Likewise, be cognizant to the use of conductivity modifiers—even doctors find it necessary at times to inject additives into the body fluids and yes, many of the patients live.

Acknowledgments

The efforts and diligence of the Electric Power Research Institute (EPRI), Exxon, Shell, Gulf, Union Electric, Consolidated Edison, Alabama Power, Massachusetts Institute of Technology, Rennselaer Polytechnic Institute, General Electric, McGraw Edison, Harley Pump Works, Doble Engineering, Westinghouse Electric, and K. D. Mills of TU Electric are sincerely appreciated and herein acknowledged.

Notice

Neither TU Electric, employees of TU Electric, nor any person acting on their behalf:

a) makes any warranty or representation, expressed or implied with respect to the accuracy, completeness, or usefulness of the information contained in this report, or that use of any information, apparatus, method or process disclosed in this report may not infringe privately owned rights,

nor

b) assumes any liabilities with respect to the use of, or for any damage resulting from the use of any information, apparatus, method, or process disclosed in this report.

References

[1] Leonard, J. T., "Pro-Static Agents in Jet Fuels," NRL Report 8021, Naval Research Laboratory, Washington, DC, August 1976.
[2] Takagi, T., "Reliability Improvement of 500 kV Large Capacity Power Transformer," International Conference on Large High Voltage Electric Systems (CIGRE), Paper 12-02, Paris, France, 30 Aug. to 7 Sept. 1978.
[3] Higaki, M., "Static Electrification and Partial Discharges Caused by Oil Flow in Forced Oil Cooled Core Type Transformers," *IEEE Transactions on Power Apparatus and Systems,* Vol. 98, No. 4, IEEE PES Winter Power Meeting, New York, NY, February 1979.
[4] Okubo, H., "Charging Tendency Measurement for Transformer Oil," Transaction A 79 051-4, IEEE Winter Power Meeting, New York, NY, February 1979.
[5] Shimizu, S., "Electrostatics in Power Transformers," *IEEE Transactions on Power Apparatus and Systems,* Vol. PAS-98, July/August 1979.
[6] Tanaka, T., "Model Approach to the Static Electrification Phenomenon Induced by the Flow of Oil in Large Power Transformers," *IEEE Transactions on Power Apparatus and Systems,* Vol. PAS-99, No. 3, Vancouver, British Columbia, Canada, July 1980.
[7] Tamura, R., "Static Electrification by Forced Oil Flow in Large Power Transformers," *IEEE Transactions on Power Apparatus and Systems,* Vol. PAS-99, No. 1, IEEE PES Summer Meeting, Vancouver, British Columbia, Canada, July 1980.
[8] Okubo, H., "Suppression of Static Electrification of Insulating Oil for Large Power Transformers," IEEE PES Winter Meeting Transaction, IEEE PES Winter Power Meeting, New York, NY, February 1982.
[9] Abuaf, N., "Forced Vaporization Cooling of HVDC Thyrister Valves," EPRI Publications EL-2710, Electrical Power Research Institute, Palo Alto, CA, October 1982.
[10] Oommen, T. V., "Electrostatic Charging Tendency of Transformer Oils," IEEE PES Winter Meeting Transaction, IEEE PES Winter Power Meeting, Dallas, TX, February 1984, IEEE, New York, NY.
[11] Crofts, D. W., "Failure of a 450 MVA, 345/138kV Autotransformer from Static Electrification," Electric Power Research Institute (Electrical Systems Division), AC/DC Transmission Substations Task Force, New Orleans, LA, May 1984, EPRI, Palo Alto, CA.
[12] Crofts, D. W., "Failure of a 450 MVA, 345/138kV Autotransformer from Static Electrification,"

1984 Transmission Substation Design and Operations Symposium Transaction, University of Texas at Arlington, September 1984.

[13] Crofts, D. W., "Static Electrification Phenomena in Power Transformers," Minutes of the Fifty-Second Annual International Conference of Doble Clients, 1985, Sec. 6-1001.

[14] Ditzler, D. A., "Failure and Repair of a Westinghouse 345/138kV Autotransformer," Minutes of the Fifty-Second Annual International Conference of Doble Clients, 1985, Sec. 6-1101.

[15] "Acoustic Emission Detection of Partial Discharges in Power Transformers," EPRI Project RP-426-1, final report, EPRI, Palo Alto, CA, August 1985.

[16] Crofts, D. W., "An Update on the Static Electrification Phenomena in Power Transformers," The Pennsylvania Electric Association, Hershey, PA, September 1985.

[17] "Effects of BTA on Transformer Materials," Mitsubishi Electric Corp., October 1985.

[18] Brunson, P. W., Jr., "Investigation of the Failure of a 500-230kV, 480 MVA Westinghouse Shell-Type Power Transformer," Minutes of the Fifty-Third Annual International Conference of Doble Clients, 1986, Sec. 6-1101.

[19] Crofts, D. W., "Methods of Controlling Static Electrification in Large Power Transformers," Minutes of the Fifty-Third Annual International Conference of Doble Clients, 1986, Sec. 6-1001.

[20] Gasworth, S. M., "Electrification Problems Resulting from Liquid Dielectric Flow," EPRI Publication EL-4501, EPRI, Palo Alto, CA, April 1986.

[21] Crofts, D. W., "The Static Electrification Phenomena in Power Transformers," 1986 IEEE Conference on Electrical Insulation and Dielectrics Phenomena, King of Prussia, PA, November 1986.

[22] Crofts, D. W., "Transformer Characteristics that Influence Static Electrification Development," EPRI Workshop on Static Electrification in Transformers, Monterey, CA, November 1986, EPRI, Palo Alto, CA.

[23] Nelson, J. K., "Electrokinetic Phenomena in Transformer Oil," EPRI Workshop on Static Electrification in Transformers, Monterey, CA, November 1986, EPRI, Palo Alto, CA.

[24] Oommen, T. V., "Static Electrification Control in Power Transformers," EPRI Workshop on Static Electrification in Transformers, Monterey, CA, November 1986, EPRI, Palo Alto, CA.

[25] Zahn, M., "Flow Induced Electrification in Transformer Oil/Cellulosic Systems," EPRI Workshop on Static Electrification in Transformers, Monterey, CA, November 1986, EPRI, Palo Alto, CA.

[26] Damsky, B. L., "Design, Installation, and Initial Operating Experience of an Advanced Thyrister Valve," IEEE 87 WM 184-5, February 1987, IEEE, New York, NY.

[27] Brunson, P. W., "The Second Failure of a 400 MVA, 500-230 kV Shell-Type Auto Transformer at South Bessemer Transmission Substation," Minutes of the Fifty-Fourth Annual International Conference of Doble Clients, 1987, Sec. 6-1401.

[28] Hochart, B., "Oil Charging Tendency Effect on Large Power Transformers," Minutes of the Fifty-Fourth Annual International Conference of Doble Clients, 1987, Sec. 6-1501.

[29] Lindgren, S., "Progress in the Control of Static Electrification in Power Transformers," Minutes of the Fifty-Fourth Annual International Conference of Doble Clients, 1987, Sec. 6-1601.

[30] McNutt, W., "Discussion of the B. Hochart and J. P. Grandjeon Paper 'Oil Charging Tendency Effect on Large Power Transformers'," Minutes of the Fifty-Fourth Annual International Conference of Doble Clients, 1987, Sec. 6-1501A.

[31] Crofts, D. W., "The Static Electrification Phenomena in Power Transformers," *IEEE Transactions on Electrical Insulation,* in press.

DISCUSSION

C. L. S. Vieira[1] *(written discussion)*—Today there are some utilities that recondition the oil with thermovacuum equipment in the normal operating transformer and maybe with the pumps working. What is the potential effect of this treatment in the charge density or in the electrostatic charging of the system? Should this procedure be revised due to this new view of charging?

D. Crofts (author's closure)—This method is employed with the transformer deenergized and there is no potential for a-c power follow current. Perhaps we should be more cognizant of the electrification problem and watch pump operation in deenergized modes since d-c potential differences still can exist and allow d-c dielectric breakdowns of low-energy magnitude.

[1]CEPEL, P.O. Box 2754, Rio de Janeiro, Brazil.